MARKETING PROBLEMS OF FOREST INDUSTRY IN INDIA

Marketing Problems of Forest Industry in India

Indra Kumar
Alok Kumar

Commonwealth

Marketing Problems of Forest Industry in India

Edition : 2024

ISBN 81-7169-037-8

Published by :
COMMONWEALTH PUBLISHERS (P) LTD.
4831/24, Govind Lane,
Ansari Road, Darya Ganj,
New Delhi - 110002
Ph. : 23242541, 23272541
e-mail : commonwealthpublishers2016@gmail.com

Distributed by :
ARJUN PUBLISHING HOUSE
4831/24, Govind Lane,
Ansari Road, Darya Ganj,
New Delhi - 110002
Ph. : 23242541, 23272541
e-mail : arjunbooks2001@gmail.com

Laser Typesetting by :
Dimple Computers
Delhi

Printed at :
Balaji Digital
Delhi

Preface

This study involves an analytical and a critical appraisal of the Problems of Processing and Marketing of Forest Products in U.P. from as many perspectives as possible. As an explorative study it has been conducted with the object of finding out the ways and means of solving the problem experienced by the marketers. It is based on a comprehensive case study of Uttar Pradesh Forest Corporation. Though, the findings are basically for the State of Uttar Pradesh. Yet, it is earnestly hoped that they will be useful in resolving the problems of marketing of forest produce in any part of the country and elsewhere.

Forests produce many economically valuable products to the society. Man use the products either in direct or indirect form in day to day life. Government holds virtual monopoly of all the forest products. It is necessary under the law that the forest produce be sold at a competitive price. For achieving this end the U.P. Forest Department initially used to dispose off the forest produce by the conventional age old practice of auction of standing timber through the contractor system. Under this system, the sales were knocked down in favour of the highest bidder. The contractor system of extraction lasted for over a century but was abandoned due to its shortcomings, such as illicit fellings, exploitation of labour, harvesting on non-scientific lines etc. and was replaced by the departmental form of working through the Forest Corporation.

This study reviews the former (contractor system) and the existing (Forest Corporation) system of processing and marketing of forest produce with particular emphasis on timber (marketing of other produce have been touched upon). The study evaluates the development of the Forest Corporation and brings out the benefits and drawbacks. It evaluates the complete system of processing and marketing of the forest produce under the U.P. Forest Corporation and gives suggestions for its improvement.

The thesis is divided into eight chapter with the initial chapter (I & II) dealing with the general survey of forests of U.P. and the different patterns of exploitation of forest resources while the remaining chapter deal with the U.P. Forest Corporation, history of its emergence, development and the specialised functions performed by it.

The chapter I deals with the Broad Spectrum survey of Forests and Forest-Based Industries in U.P. It traces the history of forests and the way they were being utilized from the ancient period to this date. The chapter has further been divided into four subsections, viz. Section (a) - deals with the significance of Forests; Section (b) - explains the topography of the State of U.P. ; Section (c) - deals with Forests of U.P. in a detailed manner, giving the complete description of types of forest, values of forest produce, etc. while section (d) - incorporates some main industries of U.P. which are based on this important resource.

Chapter II of the thesis deals with the Former Pattern of working i.e. the contractors system. This chapter includes the study of the development of scientific management from the early times and also studies the different methods of marketing of forest produce which were present in the early years. It critically evaluates the different methods of marketing in practice and finally discusses the most prevalent method of marketing of forest produce in U.P. i.e. the contractors system. A detailed description of this method of sale is studied alongwith its drawbacks, in this chapter.

The Emergence of U.P. Forest Corporation is discussed in Chapter III. It includes the suggestions of Food and Agriculture Organization (FAO) on Production Forestry and the recommendations of the National Commission on Agriculture towards the adoption of more business like approach (commercial approach) to forestry development. This chapter also gives a brief idea of Forest Corporation of different States and explains the unlike nature of U.P. Forest Corporation. It also studies the reasons of the creation of the U.P. Forest Corporation and discusses the Corporation in detail, alongwith its progress.

Aims, Objectives and Functions of U.P. Forest Corporation are dealt in Chapter - IV. This chapter includes the discussion of the various aims for which the Corporation has been formed and the various objectives which it will fulfill and also deals with an evaluation of its achievements. Chapter V solely deals with the changing organizational structure of the U.P Forest Corporation from time to time to cope up with the increasing responsibilities entrusted to it. The chapter gives as concise description of the principles of organization and explains the changing organizational structure (with the help of organizational chart) of the U.P. Forest Corporation from the year of inception to the present day i.e. 1985 - 86.

Chapter VI deals with the Processing System under U.P. Forest Corporation. The complete system of processing i.e. extraction from the point of sale of forest produce by the Forest Department to the Forest Corporation on royalty to the time when the extracted forest produces reaches the Corporation sales-depot is studied in detail. The chapter includes the systematic study of felling by various method, conversion, and transportation of forest produce from forest coupes to sales-depots. The fixation of royalty, its payment, rules an condition of working, labour, etc. and finally an evaluation of complete processing system is discussed in this chapter.

The chapter VII of the thesis discusses the Marketing of Forest Produce by U.P. Forest Corporation. This chapter deals with the concepts of marketing and the relevance of marketing in forestry. It studies the complete marketing procedure of the Corporation from the time of entry of produce in the depot to the time of export of produce from the depot in a systematic manner. The chapter also includes the study of different methods of sale, i.e. auction, private treaty, retail-sale, etc. as adopted by the Corporation. It also studies the terms and conditions of sale and payment and progress of sale from the year of inception to the present date. The evaluation of marketing has also been studied.

The last chapter (i.e. Chapter VIII) of the thesis deals with retrospects and prospects. This chapter gives the complete appraisal of the Corporation. This chapter evaluates the Forest Corporation and gives suggestion for future development purposes. In this study, all opinions expressed and conclusions reached are my own and do not reflect the views of the Government.

The thesis is based on the results of personal investigations, field visits (visits to forests), visits to sales-depot and contact with those who are in the Forest Department and Forest Corporation. The work had been beset with difficulties created by the suspicious and conservative nature of most forest employees, who were not ready to part with the information. But when they were assured of the bonafide object of my investigations, they responded to my querries. I have endeavoured to have corroborated evidence, and applied checks, and counter-checks before I ventured to adopt them in my own-ways. I have made full and free use of every possible source of information with acknowledgements. I have marshalled and presented the data, thus collected in as comprehensive and precise a manner as was possible.

(ALOK KUMAR)

Acknowledgement

I am grateful to the Government of India, the Government of U.P. and U.P. Forest Department for the use of various reports, statistical data and publications. I am indebted to Forest Research Institute and College, Dehradun and Inidan Institute of Forest Management, Bhopal for permitting me to use their libraries, various teaching cases, notes and monographs. I am also grateful to the U.P. Forest Corporation for making me available the Act of the Corporation, Yearly Reports, etc. for use and for permitting me to visit the working sites and sales-depots.

It is with a feeling of sincere regards the I hereby express my grateful thanks to my Guide Dr. Indra Kumar, Prof. of Commerce. I am deeply indepted to him for his encouragement guidance, valuable suggestions and preferred advice at every stage of this work.

I offer my heartfelt gratitude to Prof. G.M. Ram, Principal, St. John's College, Agra for being the source of inspiration. I thank him for his constant encouragement and personal interest in the work. In also thank him for providing me all possible help and facilities.

I am grateful to all my all friend, who were good and kind enough to have helped me in my work. I shall not try to name them all, however, I would like to thank Mr. Day, Regional Manager (Lucknow), U.P. Forest Corporation, who explained and discussed the various facets of the Corporation. I sincerely thank him for providing ready information and valuable suggestions from time to time. I thank Mr. Kaushalendra Prasad, I.F.S. Conservator, Utilization Circle, Former General Manger, U.P. Forest Corporation for his valuable suggestions and criticism.

I take this opportunity to convey my sincere thanks to my friends Mr. Naresh Sinha, Mr. Josiah Simon and Miss Shobha Sharma for their generous help and cooperation. To my parents, whose forebearance and understanding have been quite overwhelming, I owe a sense of dedication.

In the end I wish the thank the field staff of the U.P. Forest Corporation and the U.P. Forest Department for providing me with practical information, and Mr. Mahendra Singh for typing the manuscript so accurately in a short time.

Contents

CHAPTER ONE

A Broad Spectrum Survey of Forests And Forest Industries In U.P.

Forests existed much before the existence of human race on our planet. The ancestors of homo-sapiens depended for their subsistence upon forests and wild animals. Gradually came the change and man started living in organised communities and civilised societies, this profoundly renovated the relationship between him and the forests. The rising population and gregarious habits of the human race accounted for large forest areas being cleared of vegetation for habitation and cultivation. Forests in abundance had one kind of problem while its decadence created a new problem, though both the problems finally resolve themselves into the problem of survival of man on earth.[1]

Way back till the sixties of last century, the forests were considered no man's property; whence everybody was free to cut what he liked and whatever, way he liked. The tigers, the bears, the elephants and other dangerous denizens of forests provided the sole deterrent to the wanton destruction of their domain. General lack of law and order, meagre means of transport and want of communication proved a veritable blessing in disguise and secured a measure of unconscious protection to the forest. The age of primeval plenty did not last long and soon with the advent of British rule on the scene led to an unprecedented assault on the forests in quest of railway sleepers, timber for defence purposes and in search of land for large scale farming. Besides all this, vast forests were signed away to enterprising speculators, which proved a veritable danger to the existence of mankind just as abundance of forests had contributed a check to the progress of human race in its earliest phase.

The progress and advancement from the dark ages through a vast space of time down to the age of urbanization and industrialization shrank the forest to an extent that they became scarce. This scarcity was caused due to the wanton destruction by ever-increasing population and the establishment of new factories, railways, workshops alongwith urbanised societies. The havoc wrought in the forests and the dwindling supplies focussed the attention of the Britishers and proved to be the proverbial gloom before the dawn. It was in this wake of disillusionment that followed that the government launched a policy of 'Forest Conservation' - a policy that has formed the key-note eversince.

The commencement of a new era for forestry is marked by the year 1855 as it was in this year that Lord Dalhousie, the then Governor-General of India, laid down the future policy for India in a memorable announcement.[2] Dr.Dietrich Brandis, a German Forester was appointed as the First Inspector General of Forests to the Government of India in 1864 and his appointment marks the dawn of Scientific Forestry in India. It was the result of his initiative and drive, that regular Forest Department came into existence. As a step towards regulation and control, the Government of India passed Act VII of 1865, for the management of Government Forests. This act was replaced by an elaborate Forest Act XVI of 1878, which provided for the constitution of 'reserved' and 'protected' forests and later on it was substituted by the Indian Forest Act XVI of 1927.[3]

With the dawn of Independence, Forestry in India went through a sacrificial fire emerging as a dedicated servant of the people at large. For its role in the preservation of soil fertility, in the regulation of water regime and in the moderating influences on climate, forestry earned recognition of its right to land, as a foster-mother of agriculture. With the seal of approval of its aims and functions by democratic government, forestry can now look forward to a future as bright as its past.

The National Forest Policy Amendment of 1952 was done in the older forest policy. This provides that the national interests are not to be sacrificed in meeting demands of local population. It also states that the forests should cover 1/3rd of the total geographical land area, but the statistics prevalent shows that the forests of India do not conform to this scientific criterion laid by the experts. Forests of India cover only 22.75% of the total geographical area. The following table reveals that the position of forests is still worse in certain States/Union Territories of India.

It is evident from the table that only six States, viz Assam, Himachal Pradesh, Madhya Pradesh, Orissa, Sikkim and Tripura but all the Union Territories except Delhi conform to the scientific criterion laid by experts. There are certain states like Punjab, Haryana and Jammu & Kashmir which don't even have 10% of the geographical area of the State under forest, among the States, Tripura and Andaman-Nicobar Island among the Union Territories have the highest percentage of forest area to their geographical area. The State of U.P. (the State on which the complete study is based) has only 17.24 per cent of forest area to geographical area, which is almost half of the scientifically prescribed minimum. Thus, it becomes imperative that the forest area in Uttar Pradesh should be increased and proper conservation and afforestation should be done to meet the scientific criterion and attain the minimum prescribed percentage.

Forests help in the development of large number of forest based industries, viz. match, plywood, sports goods, furniture, etc. These industries are located throughout the State of Uttar Pradesh and are significant for providing large variety of items and also in generating employment.

I.(a) SIGNIFICANCE AND ECOLOGICAL BALANCE

From the early phase of our civilization during Indus Valley and Vedic times, when man adopted pastoral and agricultural way of life, he was still in intimate association with his sylvan surroundings. The lure of forest was strong enough to impart a characteristic colour to his principal institutions. A study of Vedas reveals that the three stages viz. Brahmacharya, Van-Prastha and Sanyasa were mostly spent in forests. The Vedic era was followed by epic age represented in literature by the Ramayana and the Mahabharata. During this age, forests were regarded as abode of peace where hermits, far from the din and noise of cities, persued their meditation and imparted knowledge to the desirous and inquisitive. The works of literature and art of the Maurya and Gupta

Table-1

STATE-WISE FOREST AREA VIS-A-VIS GEOGRAPHICAL AREA.

	AREA IN THOUSAND HECTARES		
State/Union Territory.	*Geographical area*	*Forest area*	*% of Forests to Geographical area*
1.	2.	3.	4.
I. STATES			
Andhra Pradesh	27,681	6,437	23.25
Assam	7,852	3,081	39.24
Bihar	17,388	2,928	16.84
Gujarat	19,598	1,959	10.00
Haryana	4,422	146	3.30
Himachal Pradesh	5,567	2,117	39.11
Jammu & Kashmir	22,224	2,104	9.47
Karnataka	19,177	3,757	19.59
Kerala	3,887	1,133	29.15
Madhya Pradesh	44,284	16,239	36.67
Maharashtra	30,776	6,672	21.68
Manipur	2,236	602	26.92
Meghalaya	2,249	637	28.32
Nagaland	1,653	289	17.48
Orissa	15,578	6,793	43.61
Punjab	5,036	206	4.09
Rajasthan	34,221	3,461	10.11
Sikkim	730	282	38.63
Tamil Nadu	13,007	2,254	17.33
Tripura	1,048	603	57.54
Uttar Pradesh	29,441	5,075	17.24
West Bengal	8,785	1,183	13.47
II.UNION TERRITORIES			
A & N Island	829	746.	89.99
Arunachal Pradesh	8,358	5,154	61.67
Dadar & Nagar Haveli	49	20	40.82
Delhi	149	4	2.68
Goa, Daman & Diu	381	131	34.38
Mizoram	2,109	713	33.81
Others	63	-	
All India	3,28,779	74,786	22.75

Source- India's Forests (1980), Central Forestry Commission, Ministry of Agriculture, Government of India.

periods still depict the forests maninly as 'Tapovana', i.e. places of meditation and worship.[4]

The medieval period of Indian history sees the emphasis of forests shifted from their being seats of learning and meditation to their being hunting grounds. Both Babur and Jehangir were keen sportsmen and took pleasure in hunting. The early years of Bristish rule did not see any change in the attitudes towards forests.It was in the middle of the nineteenth century that the importance of forests was understood, making imperative the necessity of protecting the hermitage of nature from the rapacity of man.

Besides their widespread occurrence, the forests greatly influence all sorts of life on this planet. Forests are playing a very important role in some form of the other i.e. from the day to day requirements to the entire development of the nation. Forests and Forest products have been so intimately associated with man and his activities that it is difficult to isolate them for an objective study. Early man obtained nearly everything from the forests -food, housing, clothing, medicines, fire-woods, tools, and other equipments and weapons.

Forests are an important geographical feature on this earth. Their tempering effect and moderating influence on the climate, soil erosion and floods is very pronounced. Forests serve as a natural screen against winds and storms. Through their mechanical obstruction, the velocity of air is slowed down and the current is deflected upwards. Such comparative from the inclemencies of wind movement within the forests and its neighbourhood. brings relief to both men and animals and also proves beneficial to the production of crops.[5] To counteract the havoc wrought by the winds of the Rajasthan desert, large scale afforestation is being done upon the fertile fields of western districts of Uttar Pradesh.

The vegetal cover of a forest greatly influences the temperature of soil and air within the forest. Forest air, as a rule is moister than the open air. Humidity which varies with temperature and the amount of water vapour in the air, is invariably very much higher in the forest. During the hot weather the maximum temperature at the soil surface is reduced and tender plants are generally protected from high temperature. During the winter season, the soil freezing is generally delayed and rendered less effective. Forests play an important role as 'rain-maker'. This influence of forest is due to the fact that forest trees and plants act as a pump, moving water from the soil to the air in great quantities.[6]

The forest cover decreases the amount of water evaporated from the soil. Large quantities of decomposing leaves, fruits, branches and twigs, under normal conditions, enrich the forest soil with a great deal of humus. The humus gradually mixes with the mineral soil beneath and also supplies it with soluble compounds carried downward with percolating rain water. Thus, the forest cover constantly improves the chemical and physical characteristic of soil. The combined result of interception of precipitation and retention of water by the forest floor is that the forests greatly help in decreasing amount of surplus free moisture in the soil. On the other hand, particularly in dried areas, forests also increase the amount of soil moisture by reducing surface run-off, by increasing the permeability of the soil and decreasing evaporation from it. In the forest localities, the surface run-off is slowed down by the litter and this provides more time for the water to be absorbed by the humus, which acts as a huge sponge. The humus prevents mudding of the underlying mineral soil, which remains porous, and thus, absorbs and retains much more water than the exposed soil in the open. As a result of fall in surface run-off, the soil erosion is substantially checked and stream flows are modified. The root system of the trees and plants have a binding effect on the soil and thus, save it from being washed away along the banks of streams; they also check the occurrence of land slides. The devastating power of wind as an agent of soil erosion is reduced within a forest area. This beneficial influence of forest vegetation is more effective in localities where the danger is more prominent such as on steep slopes, in light soil and with heavy rainfall[7].

The effect of forests in checking soil erosion and controlling the flow of streams has an important effect in curbing flood-peak and flood damage. Streams passing over steep slopes and deforested water shed, as a rule, carry several times much more solid material than water. Even streams with small gradients may carry enormous loads of sediments.

The forests are a perennial source of a large variety of raw-material necessary for industry. These material benefits derived from the forests are obviously more palpable to us than the indirect benefits derived from forests. A large number of industries, both large and small are dependent upon the various products drawn from the forests. These industries in turn provide large scale employment, besides being source of furnishing large (wide) variety of indispensable consumer goods.

Forests have great economic influence as they provide large quantity of wood, which can be put to diverse uses. The wood extracted serve as a multipurpose raw-material. Until the advent of oil and gas, wood constituted men's chief source of fuel and it still serves this purpose. The various uses of wood can be categorised as structural, decorative and specialized. A good deal of wood in the form of timber (structural timber) is used for the construction of buildings, bridges, derricks, scaffoldings, jetties, etc. of all kinds. The decorative features are mainly due to colour, luster, grain, anatomical structures and abnormalities. Highly figured stocks of some woods is used for high class furniture, decorative paneling, parquet flooring, fine carving and ornamental articles like antique boxes, cigar cases, plaque and curios. Decorative woods are also increasing in demand as face veneers for ornamental plywood , linen boards, flush doors, etc. The research has shown that waste wood is used for decorative purposes, like: 'the F.I.R.Diaper','Mosaic Boards' and mats.[8]

The structural uses of wood are in the manufacture of agricultural implements, aircrafts industry, bearing materials, crates and packing cases, brushes and brooms, carts and carriages, co-operage, marine piles and harbour works, match boxes, splints mathematical instruments, mine works, pit-props, motor lorry and bus bodies, musical instruments, pencils and pen-holders, railway sleepers, carriages and wagons, shingles (wood tiles), ships and boats, sports equipments, toys, bobbins, textile mills, flooring etc.[9] In fact, it can well be said that at no stage between the cradle and the coffin, can man do without the use of wood. This is the reason why the following notice is seen in the wood-lands and parks in Portugal:

> To the Way fares-
> Ye who pass by and would raise your hand against me, hearken ere you harm me.
> I am the heat of your hearth on the cold winter nights, the friendly shade shielding you from the summer sun, and my fruits are refreshing draught quenching your thirst as you journey on.

I am beam that holds your house, the board of your table, the bed on which you lie and the timber that builds your boat.

I am handle of your hoe,. the door of your homestead, the wood of your cradle and the shell of your coffin.

Ye who pass by, listen to my prayer and harm me not.

The Sphere, May 20th, 1939.

In recent years, more spectacular advances have taken place in the utilization of woody material. It has its uses in the manufacture of all kinds of paper products, the demand of which seems unlimited. From 'dissolving pulp' comes cellophane, rayon, artificial wool and many plastics. There are several other chemical substances derived from wood like wood-gas, potash charcoal, acetone, ethyl alcohol, methyl alcohol, sugar, yeast, vanillin, etc. Besides, trees and various raw material derived from them, forests are also significant for providing vast variety of minor products such as bamboos, grasses, fibres, flosses, gums, katha, barks, tannins, essential oils, minerals wax, honey, fruits, etc. [10] Forests also provide medicinal herbs, which are very important for curative purposes of varied diseases.[11]

Besides their geographic and economic significance, the forests are important from aesthetic and social points of view. The forests help in preserving a variety of fauna, which consists of galaxy of magnificient animals, colourful birds and sparking fishes. Most wild animals, birds and fish become, in part, product of the forests. Many fur-bearing animals, prized trophies, such as of tigers, panthers, bears, etc. sustain other industries. Uses of forests and its wild life for recreational purposes and for hunting help in setting up side industries like automobiles, cameras, photographic, ammunitions, clothing, insect repellents and gadgets of all sorts. Wild-life constitutes an important attraction in promoting tourism and thus, bringing foreign exchange to the country. Forests are significant from literary point also, as they provide inspiration to the poets, artists and philosophers. Forests are the symbol of uprightness, chivalry and devotion alongwith its economic, geographic and social significance.

Thus, it can rightly be said:-

Trees-grass-unique in the plant kingdom symbols of uprightness, chivalry and devotion, inspiration of poets, artists and philosophers.

Writing material of the ancients;

the styles of contemporaries.

Food, shelter and clothing of the people

Industrial substance of a thousand uses

Redeermer of waste places -protector of the soil.

I. (b) TOPOGRAPHY

The state of Uttar Pradesh is situated between 77.5° E and 85° E longitudes and 23° N and 31.3°N latitudes. [12] The northern boundary of the state extends upto the Tibetan border and Nepal along the Himalayan ranges, while the southern part of it extends upto Madhya Pradesh. Its western and eastern boundaries touch the boundaries of its neighbouring states, viz. Rajasthan, Delhi, Haryana, and Bihar respectively. In the north-western direction, it extends upto Himachal Pradesh. The fertile gangetic basin is sandwitched between the Vindhya Hills-cum-Central Indian Plateau and the Himalayan Ranges.

This unique situation of the state provides it with great variety of topographic features. The main physiographic divisions of the state are:

(A) THE HIMALAYAS

These ranges run more or less parallel to each other from north-west to south-east direction intersected by low valleys to deep gorges, cut by many rivers of which Ganga, Yamuna and Sharda are notable. The elevation varies considerably reaching a maximum of 7,817 metres at Nanda Devi.

(B) SUB-HIMALAYAS

The sub-himalayan division is also known as Tarai-Bhabar. This narrow belt comprises the lower foot hills of Himalayan and Shivalik formation and its adjoining fringe where rivers debouch into the plains. It is divisible into Bhabhar and Tarai portions having low and high water tables respectively.

(C) GANGETIC PLAIN

This is an extensive and more or less flat plain traversed by many rivers.

(D) CENTRAL INDIAN TABLE LAND

The south-western portion of the state mainly comprising Jhansi and part of Banda districts forms a portion of the Central Indian Table land, constituting the catchments of rivers like Chambal, Betwa and Son which run from South to North and then join the Yamuna and the Ganga.

(E) VINDHYAN TRACT

The Vindhyan tract is mainly confirmed to Varanasi and part of Banda and Mirzapur districts in the southern part of the State.

II. CLIMATE

The extensive variation in the topographic features of the State gives birth to diverse climate conditions. There is cold montane climate as in Himalayas especially above the snow-line where the temperature remains below the freazing point, while there are also excessively hot and arid regions in the southern plains, especially adjoining the Thar desert where mercury goes upto 45.5° C during the summer months. Baring the hill stations and Dehradun situated in the foot hills, the variation of temperature is not much. On the average Agra and Jhansi have the highest temperature and Bareilly and Roorkee the lowest.

In general, the climate of the plains may be described as monsoonic during the summer it is dry and hot and in the winter, it is mild and frost free. The greatest heat develops before the onset of the rainy season, although after the rainy season is over, the temperature in autumn again rises slightly and the humidity does not fall proportionately. The highest amount of precipitation takes place during the rainy season and winters are not usually damp. The northern portion of state comprising the hilly tracts have a montane climate where the decisive factor is the influence of the altitude and aspect.

The monsoon breaks towards the end of June and continues till September. A little rainfall also occurs in winter months due to north-eastern monsoon. The areas above 1835m altitude have snow fall during the winter. Maximum precipitation occurs in the outer hills. Relative humidity is high at Gorakhpur, slightly lower at Allahabad, Lucknow and Meerut, and lower still at Jhansi and Agra. Rainfall varies considerably throughout the State, the eastern and submountaineous train receiving a slightly higher rainfall and the south-western part, the lowest rainfall. The rainfall is less then 10" in the sub-alphine region and over 100" along the foot hills. The high precipitation in the outer hills and the tarai belts support most of the luxuriant forests of the State.

The great variation in the climate condition can be classified under the following main climatic zones:

(i) Alpine

In the extreme north is the narrow belt lying above 3340 metres, where annual precipitation is less than 25 cm., most of which falls during winter as snow. The climate is conditioned by the altitude and aspect.

(ii) Montane Temperature

In it also altitude is again the decisive factor. Precipitation varies from 25 cm to 100 cm of which about 30 per cent falls as snow during the winter and the rest is received as rains during the monsoon period. The winters are rather long and pronounced and summer generally mild.

(iii) Sub-Tropical

The low level valleys extending over considerable distance into the interior with intervening ridges ranging from 1700 to 3350 metres comprise this region. The annual precipitation varies with local topography. Fifteen per cent of precipitation is received as snow and 75 per cent as rains. The southern sphere of main himalayan range receive the highest precipitation which may be 200 cm or more, whereas in the inner, sheltered valleys the rainfall is about 100 cm. The summer temperature in low valleys is very high.

(iv) Moist-Tropical

The sub-himalayan region together with the eastern gangetic plain receive an annual precipitation of 100 cm or more mostly during the monsoon. The summer temperatures are very high and in part of the region hot dessicating wind prevails.

(v) Dry-Tropical

The western part of the State receives an annual rainfall of less than 100 cm. The summer temperture is more than 45° C and hot dessicating wind prevails as a rule. The winter is more severe than in the eastern region, with occasional frost.

III GEOLOGY

Uttar Pradesh can broadly be divided into five geological regions:

(i) The Himalayan Tract

The Himalayan-Shivalik tract lies on the extreme north. The Himalayas originated in Eocene orogeny. This tract comprises of the hills of Tehri-Garhwal and Kumaon circles, and begins from the snowy ranges. There is a south-wardly recession of hills of decreasing heights, until they merge with Shivalik formation. In the inner ranges of the Himalayan massif, granites and quartzites predominate, giving place to geneisses, biotite schists and phyllites in the outer spurs.

The Shiwalik range of hills of upper tertiary age separate Dehradun from the plains and fringe the hills of the southern foot of the himalayan ranges. The lower portion of the system is characterised by a great thickness of fine-grained, grey, micaceous, sandstone, interbedded with clay bards near its lower portion; while the upper part is composed of soft earthy clays, undistinguishable from the alluvium of the plains except by the disturbance they have undergone, and the presence of coarse conglomerates of well founded pebbles and boulders of crystalline and metamorphic rocks of himalayan region.

(ii) The Sub-Himalayan Tract

The sub-himalayan tract consists of three distinct proportions viz.: (a) Bhabhar area immediately below the hills, (b) Tarai, and (c) The abutting plains strip. This tract clings to the lower foot of the Shiwalik. The Piedmont terraces of gravel along the foot of the Himalayas; called the Bhabhar consists of thick layers of boulders in a matrix of sand and gravel, and covered by a mantle of soil. The streams, as they emerge from the hills, percolate through and sink into the gravel. The Bhabhar is, therefore, a waterless tract. It is covered with high forests in which Sal (Shorea robusta) predominates. At the base of the slope, much of the water which has percolated the gravel reemerges in the form of springs. The ground is marshy and high grasses replace the forests. The tract is known as Tarai.

(iii) The Gangetic Plains

The gangetic plain is the most populous and rich portion of the state and it lies on the immense alluvial plain of the river Ganga and its tributaries. The general height of the surface is not much but the landscape has gradual gradient from north-west to south-east. The alluvial deposits are of great thickness and comprise of successive layers of sand, sandy silt and clay. The older deposits occupy the higher land and are liable to denudation. The characteristic ravine land occurs in these tracts.

The old alluvial consists of massive beds of pale reddish brown colour which at times become yellowish or even ash-grey. While in some places pisolitic concretions of hydrated iron-oxide in small pieces may be found disseminated in the soil beds, in places Kanker forms compact beds of indurated character. Sand, gravel and conglomerates also occur. The newer deposits generally occur in the neighbourhood of rivers and consists of coarse gravel near the hills, sandy clay and sand along the course of rivers and fine silt in the flatter parts of the plains.

(iv) The Bundelkhand Tract

The Central Indian Plateau is also known as the Bundel khand tract. The common rock of lower Bundelkhand is gneiss which consists of six minerals- red orthoclase felopar, white plagioclastic felopar, quartz, horn blende, chlorite and mica. In the north-east, the gneiss is covered with superficial deposits, elsewhere by overlying Vindhyan or other formation. The land is undulating or flat, with a few hills. In the lower Bundelkhand, however, long narrow serrated ridges composed of quartz reef form a most striking feature of the landscape, running in straight lines generally with a north-east to south-west direction and sometimes attaining a height of 200 metres.

(v) The Vindhyan Tract

The geological formation occupies part of Varanasi and the major portion of Mirzapur district. The upper Vindhyan occur in the north, the Kaimur ranges in the centre and the lower Vindhyan rocks in the southern portion of the district. The lower Vindhyan occurring south of river Son consists of the Bijawar shales and slates. Further, to the south occur granitic and gneissic rocks of great age. In greneral, the lower Vindhyans are argillacious and calcareous and the upper arenaceous and argillacious and there is a sudden and widespread change from the fine grained deposits of the Rohtas to the coarse sandstone of the Kaimur group.

IV. SOIL

The great variation in the climatic condition and the variety of geological formation reflects the variation in the types of the soil. These numerous types of soil can be classified as under:

(i) Himalayan Hill Soil

The soils in the hills can be differentiated into four distinct genetic types: (a) The red loams are mostly found along the slopes of lower hills or along ridges and the profiles are generally sandy in nature. (b) The brown forest soil is in higher sub-tropical and lower temperate regions. (c) The podsol is met with under mild slope gradients and in pockets of hills and ridges and in shady places in both sub-tropical and temperate zones. (d) The weisenboden or meadow soil is found near water streams and cold and low lying shady places.

(ii) Bundelkhand Soil

This is found to the south-west of river Yamuna and are of two types, viz. : the black soil and the red soil. The black soil is found under restricted drainage and occurs in low lying and flat tracts and valley bottoms with impeded drainage. The red soil occurs on plateau tops and upper slopes and is gradually well drained.

(iii) Bhabhar and Tarai Soil

This type is found in the sub-montane tract at the foot of the himalayas which is in the north of the State from Dehradun to Deoria. Five textural-cum-calcimorphic soil associations, varying from clay-loam to sandy-loam can be recognised. The organic status of this soil is low.

(iv) Vindhyan Soil

The Vindhyan soil also belongs to the red and black groups and occurs in the tract covered by Mirzapur and Varanasi districts. Red soil of the Vindhyan up land occupy high level sites and are basic alluvial in nature being formed under excessive drainage. The black soil are alluvial and occur in the Vindhyan low land.

(v) Intra-Zonal Soil

The soil with more or less developed characteristic, reflecting the dominating influence of some local features of relief, parent material of age over the normal effect of climate and vegetation are classed as inter-zonal soil. They are of two types : (a) Hydromorphic -occurs in low lying tracts of the lower gangetic plains of U.P. and in the lower gangetic basin in the part lying beyond the confluence with Yamuna. (b) Halomorphic occurs in vast stretches of saline and alkaline land popularly known as usar or reh, and are lying waste and uncultivated. The districts mainly affected by this are Aligarh Mainpuri, Kanpur, Fatehpur, Unnao, Etah, Etawah, Raibareli and Lucknow.

(vi) Alluvial Soil

These soils developed on the Indo-gangetic alluvium, may be divided into three ranges as follows:

(a) *Western-Range* : The Tarai area in the north-west part of the range comprising the districts of Nainital and Dehradun consists mostly of dark-grey soils, varying from loam to sandy loam, single grained in structure and generally rich in organic matter. In the plains of western parts of the range (e.g. districts of Saharanpur, Muzaffarnagar and Meerut), the soils are deep and generally very fertile, while towards further east the tract (Bareilly, Bijnor, Pilibhit and Moradabad), contains soils which are generally heavy loam in texture and single-grained in structure.

(b) *Central Range:* The north-western part (Kheri and Sitapur) - Here the soils are loam or sandy loam and are slightly acidic in nature. In the other parts the soils are sandy loam in character. The nitrogen content is poorest in Sitapur district, average in Kheri, where as Kanpur district is the richest.

(c) *Eastern Range:* There are two types of soil locally known as bhat and banjar; a third type of soil known as dhuh occurs near banks which are liable to inundation. Bhat soils are generally low-lying single-grained and sandy loam in texture having high lime contents and high water holding capacity.

Banjar soils are generally high lying varying from loam to sandy loam, are marked by lower lime content and are slightly acidic in reaction. In general, the soil of the northern parts of Gorakhpur and Basti districts are richer in nitrogen than those in southern parts.

I. (c) FORESTS OF U.P. - A PANORAMA

The great epics, the Mahabharata and the Ramayana, mention of the vast forests over large tract of the State of Uttar Pradesh. During the medieval period, the Rajahs and Nawabs maintained the more accessible portions of the forests as games preserves for their personal enjoyment but no serious attempts were made to regulate the working of the forests. Besides, these considerable stretches were still under natural thick forests which could easily meet all needs of the people. Then came the British rule over this area, bringing unprecedented assault on the forests.

It was in the year 1935, that the forestry was transferred from centre to the Provincial Government, under the Government of India Act 1935.[13] The provincial government did their best for promoting the idea underlying the National Forest Policy of 1927. At the beginning of the last world war, the government did nothing to improve forestry, but fully occupied with the supply of timber. The forest area under the control of forest department was 13,571 sq.km (5,240 sq. miles) in 1935. This shows that the forests were inadequate and urgently needed conservation and development.

After independence in 1947, various steps of conservation, preservation and development were taken. In its wake, the U.P. Private Forest Act, 1948 was passed, which was followed by the Zamindari Abolition and land Reforms Act of 1950. [14] The merger of princely states of U.P. saw a great increase of 3537.30 sq. miles of private forest under the forest department.[2] The area doubled under the control of the States. Then came the five year plans, in which afforestation was done and large land was brought under direct control of the State, making an all out effort to rehabilitate and make the forest land productive under the forest department and also to preserve the existing tree wealth that may be in the possession of others. Various schemes were taken for plantation, afforestation and reforestation, thus, increasing the forest under the direct control of the U.P. Forest Department to a great extent.

The forests of U.P. are given below in Table 2:

Table 2

1.	Total Geographical Area in U.P.	-	2,94,413 sq.km.
2.	Total Forest Area in U.P.	-	51,212 sq. km.
3.	Forest Area as percentage of Geographical area	:	17.39%

These forests are classified as follows :

Forest Area Under the Control of U.P. forest Department :

(a)	Reserved	-	34, 579.20 sq.km.
(b)	Protected	-	1, 685.04 sq.km.
(c)	Unclassed and vested	-	4,368.98 sq.km.
	Total	=	40,633.22 sq. km.

Other Forests

(a) Civil Forests under the control of Revenue Authorities:

i)	Almora	-	1,821.00Sq.km.
ii)	Pithoragarh	-	1,210.00sq.km.
iii)	Garhwal	-	1,806.00 sq. km.

iv)	Chamoli	-	1,043.00 sq. km.
v)	Nainital	-	192.00 sq. km.
	Total	=	6,072.00 sq. km.

(b) Soyam Forests

i)	Uttar Kashi	-	147.89 sq.km.
ii)	Tehri	-	1277.74 sq. km.
iii)	Dehradun	-	516.00 sq. km.
	Total	=	1941.63 sq. km.

(c) Panchayat Forests

i)	Almora	-	627.99 sq.km.
ii)	Garhwal	-	296.40 sq. km.
iii)	Nainital	-	207.67 sq. km.
iv)	Chamoli	-	521.49 sq. km.
v)	Pithoragarh	-	714.45 sq. km.
	Total	=	2368.00 sq.km.

d) Private Forests

i)	Dehradun	-	158.36 sq.km.
ii)	Almora	-	0.52 sq. km.
	Total	=	158.88 sq. km.

(e) Municipal and Cantonment Forests

i)	Dehradun	-	10.82 sq.km.
ii)	Nainital	-	7.42 sq.km.
iii)	Garhwal	-	6.07 sq. km.
	Total	=	24.31 sq. km.

(f) Other Forests

i)	Nainital	-	14.53 sq. km.
	Grand Total	=	51,212.57 sq. km.

SOURCE - Forest Statistics (1985) U.P., pp. 1-3, U.P. Forest Department, Government of U.P.

The forests under the department don't resile with scientific distribution of 60 per cent in the hills and 40 per cent in the plains.

Instead of this, they occupy 23, 755.15 sq. km. in the hills and only 17,050.40 sq. km. in the plains. Thus, the forests under the control of department being 47 per cent in the hills and 7 per cent in the plains, to the geographical area of the hills and plains respectively. Besides this, the (Dept.) Government also has control on the roadside avenue and canal banks. The road-side avenue has shown an increase from 14, 859 km, in the 1976 to 19,903 km. in 1979. This further increased to 23,130 km in 1982, while the area along the canal bank remained the same i.e. 2162 km. A complete table showing the total area under road-side avenue and canal bank is given in table 3:

Table-3

ROAD-SIDE AVENUE AND CANAL BANKS UNDER THE CONTROL OF FOREST DEPARTMENT.

	AS ON			
	31.3.79	31.3.80	31.3.81	31.3.82
Road-side avenue	19903 km	20995 km	22334 km	23130 km.
Canal Banks	2162 km	2162 km	2163 km	2162 km.

SOURCE : Own compilation on the basis of Annual Report of U.P. Forest Department.

These forest areas are not equiproductive and have wide variation in their economic exploitation. That is why they have been categorised as follows from the point of economic utility :

Thus, the above table shows that of the total Forest area only 24,022.27 sq.km is exploitable forests and 2,599.03 sq.km. is unworkable forest while 1,859.95 sq.km. is the area containing inferior forests. Beside this, the remaining area is either Blank or Vacant, which calls for its development by species of economic value.

TYPES OF FORESTS

The forests show diversity in composition because of their wide geographical and altitudinal distribution. They extend from the snow clad mountains in the north to the Vindhyan Plateau in the south. The woody plants occurring in the State add upto a thousand species. Even, if species which are sometimes herbaceous are excluded, there are more

Table-4

FOREST AREAS UNDER DIFFERENT CATEGORIES.

		Square Kilometers :
(a) Exploitable Forests	-	24,022.27
(b) Unworkable Forests	-	2,599.03
(c) Waste lands and Inferior Forests:		
i) Area containing inferior type of forests where enrichment by plantation is required		1,859.95
ii) Vacant areas-		
-Already planted by forest department	:	2.203.39
-Plantable areas	-	1,809.73
-Usars	-	278.67
-Water-logged	-	87.97
-Sandy, river-beds, boulder, etc.	-	1,071.36
(d) Blanks:		
i) Grass land	-	1,647.16
ii) Degraded	-	542.39
iii) Snow covered	-	2,947.94
(e) Others	-	1,617.32
Grand Total	=	40,687.18

SOURCE: U.P. Forest Statistics (1981-82), p.5, U.P. Forest Department, Govt. of U.P.

than 300 trees, about 400 shrubs and over a 100 woody climbers. More than 220 species of grasses have been collected from the gangetic plain alone and many important species of medicinal plants like Rauwolfia Serpentina, Viola serpens, Podophyllum hexendrum, Ephedr a gerardiana, to name only a few, are represented. In the hills, the main species are Pinus roxburghii (Chir), Abies Pindrow (Fir), Picca symthiana (Spruce), Cedrus deodara (Deodar), Cupressus torilosa (Surai) and Pinus Wallichiana (Kail). In the sub-mountain areas, the prominent species are Shorea robusta (Sal), Terminalia tomentosa (sain or asna), Cedrela toona (teen), Adina cordifolia (Haldu), Holoptelea integrifolia (Kanju), Sygyzium cuminii (Jamun), Acacia cetechu (Khair), Delbergia sissoo (Sissoo), Salmalia malabarica (Semal), Trewia nudiflora (gutel), etc. In the plains Acacia arabica (Babul) is an important species.[15]

Table - 5

Sl No	*Species*	*As on 31.3.82 (in Sq.Km.)*	*Percentage with total Forest area.*
1.	Chir	4,11,782	10.12
2.	Deodar	18,411	0.45
3.	Fir and Spruce	89,405	2.20
4.	Kail	18,417	0.45
5.	Cypress	3,295	0.08
6.	Sal	5,35.287	13.16
7.	Teak	51,792	1.27
8.	Khair	74,429	1.83
9.	Sissoo		
10.	Miscellaneous species	19,07,780	46.89
11.	Unproductive, blanks,etc.	9,58,120	23.55
	Total	40,68,718°	100.00

SOURCE: U.P.Forest Statistics, p.4, U.P.Forest Department.

The area of forest occupied by some of the important industrial species is indicated below:

From the table, it is clear that the major portion of Forest in U.P. is constituted by species like Sal (13.16%) and Chir (10.12%) while the least area of forest is covered by Cypress (0.08%). Thus, it is clear that maximum area of forest in U.P. is covered by Sal, Chir, Fir and Spruce and Khair, Sissoo and Teak which are of great industrial importance.

The forest in the state is not having identical merchantibility potential in all the areas. That is why they are classified under different categories, from the point of view of merchantibility.

From the table it is clear that the total area dedicated to timber is 26,40,208 sq.km. and in others is 14,28,510 sq.km. Of the total area dedicated to timber, 19,27,812 sq.km. is merchantable and 2,75,426 sq.km. is potentially merchantable, while 4,36;970 sq.km.is though

Table- 6

(A) MERCHANTABLE		
i) Dedicated to timer		19,27,812 sq.km.
ii) Others		6,02,802 sq.km.
Total	=	25,30,620 sq.km.
(B) POTENTIALLY MERCHANTABLE		
i) Dedicated to timber	-	2,75,426 sq.km.
ii) Other		3,03,595 sq.lm.
Total	=	5,79,021 sq. km.
(C) UNMERCHANTABLE		
i) Dedicated to timber	-	4,36,970 sq.km.
ii) Others	-	5,33,946 sq.km.
Total	=	9,70,916 sq,km.
Total area dedicated to timber	-	26,40,208 sq.km.
Total area under others	-	14,28,510 sq,km.
Grand Total	=	40,68,718 sq.km

SOURCE: Forest Statistics U.P. P.7, U.P. Forest Department.

dedicated to timber is unmerchantable. Of the total area which is under other species, 6,02,808 sq.km. is merchantable, 3,03,595 sq.km. is potentially merchantable and 5,33,946 sq.km. is unmerchantable.

The forests are having huge burden on them as they are scarce in relation to the increasing demand caused by their varied uses. They are very essential for the existence of human and bovine population. The increasing human and bovine population overburden the forests thus, creating an imperative necessity for the conservation, plantation and development alongwith exploitation of inaccessible untapped forest resources. The forests of U.P. have to fulfill the demand of nearly 1108.86 lac (Male-587.93 lac and Female- 520.93 lac) human beings and 6,03,80,000 bovine population, thus making them inadequate.

Forests provide various types of employments. A vast permanent property like the forest estate of U.P. calls for an adequate administration besides a large number of labour force. The administrative and executive officers are supported by a well organized office establishment. Further, in the field of field management (which includes work on forest portection and regeneration) collection of forest produce, fellings conversions and extraction of timber and forest maintenance provides employment to many skilled and unskilled workers, whole time or part time. Apart from this, many people find work in forest based industries (large- scale, small-scale of cottage industries).

The total employment provided by the forests can, be categorised as under:

(A) Employment in Forest Department.

(B) Employment of seasonal staff excluding labour on daily wages.

(C) Employment generated under plan scheme.

(D) Employment of labour in Forest and Forest based industries with consumption of wood and minor forest products.

The following table will reveal employment potential under various categories;

The table shows that the total regular employment in forest department was 16,510 on 1.1.83. Of which 366 were gazetted officers, 9675 were non-gazetted, executive and technical staff and 6469 were the ministerial, class IV and miscellaneous staff. This reflects that the total

Table-7

(A) EMPLOYMENT IN FOREST DEPARTMENT, U.P.

S.No.	Designation		Number
1.	Gazetted		366
2.	Non-gazetted , Executive and Technical staff		9675
3.	Ministerial, class IV & Misc. staff		
	Total	=	16510

permanent employment was quite insignificant in relation to the vast forest of the State.

The forest department employees 5,559 persons as seasonal staff in the department besides the regular employees. This seasonal staff excludes labour on daily wages.

(B) EMPLOYMENT OF SEASONED STAFF EXCLUDING LABOUR ON DAILY WAGES:

S.No.	Categories		No. of employees
1.	Stump Marker	-	1,750
2.	Plantation Jamadar/Water	-	885
3.	Fire Watcher.	-	1,210
4.	Grazing watcher	-	140
5.	Other watchers	-	231
6.	Fencing Guards/Other Guards	-	47
7.	Export Moharrir	-	252
8.	Road Moharrir	-	95
9.	Mali	-	433
10.	Miscellaneous	-	516
	Total	=	5,559

(C) EMPLOYMENT GENERATED UNDER PLAN SCHEME:

In Lakh Man Days

Sector	*V Plan 1974-79 Actual*	*1979-80 Actual*	*VI Plan 1980-85 Estimated*	*1980-81 Actual*	*1981-82 Actual*
I. State Plan :					
i) Forestry	148.75	54.41	472.98	100.42	162.36
ii) Soil Conservation	47.97	17.71	108.48	15.12	16.71
Total	196.72	72.12	581.46	115.54	179.07
II Centrally sponsored					
i) Forestry	30.78	1.52	28.98	0.95	5.57
ii) Soil conservation.	21.60	12.96	127.50	21.33	18.40
Total	52.38	14.48	156.48	22.28	23.97
Grand Total	249.10	86.60	737.94	137.82	203.04

Under the plan scheme the total employment in the Fifth Plan was 249.10 lakh man days of which 196.72 lakh man days was from State Plan and 52.38 lakh man days was from Centrally sponsored plan. It was also estimated that the total employment under the Sixth plan will increase to 737.94 lakh man days. It is also clear that the total employment under plans in 1980-81 and 1981-82 was of the order of 137.82 lakh man days and 203.04 lakh man days. This showed that the forestry sector also provided employment to quite significant population.

The 'table reflects that the total number of persons employed annually in approximately 3500 industries (300 man days=one labour) during 1977-78 was 2,19,070, which increased to 2,34,382 in 1978-79.

(D) EMPLOYMENT OF LABOUR IN FOREST AND FOREST BASED INDUSTRIES WITH CONSUMPTION OF WOOD AND MINOR FOREST PRODUCTS.

Category	No.of trial	Indus units	Labour employed in lac man days average	No. of persons employed anually /300 Mandays= 1 person
A) Production of Forest Produce:				
1977-78			657	2,19.070
1978-79			703	2,34,382
1979-80	3500	Approx.	645	2,15,000
1980-81			561	1.86,910
1981-82			621	2,07,133
B) Forest Base Industries			Included above.	

SOURCE: Forest Statistics, U.P., pp. 117-120, U.P. Forest Deptt.

In 1979-80, the labour employed fell to 2,15,000 and 1,86,910 during 1980-81. But in 1981-82 the employment increased to 2,07,133. This shows that forest based industries also create large employment to the labour force.

Besides this, the forests gives large employment to labourers on daily wages and also create employment in the other related works like construction of road in forests for extraction of wood, employment given by the contractors for cutting and bringing the timber to their depots and employment created by other schemes aided by the World Bank, etc.

The forests have always been looked upon as commercial and quasi-commercial proposition. They have always been a source of revenue owing to its economic utility for various purposes. The forest department earns is a large revenue from the sale of its both major and minor products. The following table reveals the total value of major and minor forest products respectively:

Table-8

VALUE OF MAJOR FOREST PRODUCTS IN THOUSAND RS.

Year	*Timber*	*Fuel*	*Total*
1975-76	26,21,01	1,10,63	27,31,64
1976-77	27,31,64	1,29,19	28,60,63
1977-78	33,24,26	1,23,69	34,47,95
1978-79	36,19,46	1,42,41	37,61,87
1979-80	32,77,70	1,65,14	34,42,84
1980-81	35,63,58	1,71,00	37,34,58
1981-82	43,47,85	1,96,09	45,43,94
1982-83	42,11,04	1,91,61	44,02,65
1983-84	40,35,68	1,83,85	42,19,53

SOURCE: Own compilation on the basis of Forest Statistics, U.P., U.P. Forest Department.

The total value of forest produce - major and minor, can be jointly discussed with the help of following table:

The table above gives the complete details of table 8&9. The table shows the total value of forest produce by both major and minor produce in various years. The table helps in drawing the following inferences:

Table-9

VALUE OF MINOR FOREST PRODUCE IN RS.

Year	*Value in Rs.*
1978-79	Rs.10,17,88,562/-
1979-80	Rs.11,60,10,733/-
1980-81	Rs.13,84,26,425/-
1981-82	Rs.13,55,21,305/-

SOURCE: Own compilation on the basis of Forest Statistics, U.P. U.P. Forest Department.

Table-10

FIGURE OF VALUE OF FOREST PRODUCE (IN THOUSAND RS.)

Year	*Minor forest produce*	*Fuel*	*Timber*	*Total*
1975-76	9,58,99	1,10,63	26,21,01	36,90,63
1976-77	8,93,35	1,29,19	27,31,64	37,54,18
1977-78	8,71,14	1,23,69	33,24,26	43,19,09
1978-79	10,17,89	1,42,41	36,19,46	47,79,76
1979-80	11,60,11	1,65,14	32,77,70	46,02,95
1980-81	13,84,26	1,71,00	35,63,58	51,18,84
1981-82	13,55,21	1,96,09	43,47,85	58,99,15
1982-83	18,37,33	1,91,66	42,11,04	62,40,03
1983-84	17,04,67	1,83,85	40,35,68	59,24,20

SOURCE : Own compilation on the basis of Forest Statistics, U.P., U.P. Forest Department.

a) The value of minor forest produce declined from Rs.9,58,99000/ in1975-76 toRs.8,93,35,000/-.andRs.8,71,14,000/- innext two years 1976-77and1977-78 respectively. The value since 1978-79 started rising from Rs.10,17,89,000/- and reached to Rs.13,84,26,000/- in 1980-81. In 1981-82 it showed a decline but later on it again increased and reached to Rs.18,37,33,000/- in 1982-83. While the value of minor forest produce decline to Rs. 17,04,67,000/-in 1983-84.

b) In the case of fuel, the value increased from Rs.1,10,63,000/- in 1975-76 to Rs.1,96,09,000/- in 1981-82. In the year 1982-83 and 1983-84, the value from fuel declined to Rs.1,91,66,000/- and Rs.1,83,85,000/- respectively.

c) The value of timber showed a steep rise from 1975-76 to 1978-79 i.e. from Rs.26,21,01,000/- to Rs.36,19,46,000/- and then a decline in 1979-80 to Rs.32,77,70,000/-. After 1979-80, the value of timber again increased steeply and reached the maximum of Rs.43,47,85,000/- in 1981-82. In the next two years it showed a gradual decline and reached to Rs.40,35,68,000/- in 1983-84.

d) The total value of forest produce constantly increased from 1975-76 to 1982-83 except a decline taking place in 1979-80. The total value of forest produce increased from Rs.36,90,63000/- in 1975-76 to Rs.62,40,03,000/- in 1982-83, but in 1983-84 in declined to Rs.59,24,20,000/-.

The forests though having economic importance are unable to collect the revenue equal to the total value of the forest produce. The revenue expenditure chart given vide table 11 explains the expenditure incurred by the forest department in various years:

The table 11 depicts the following:

(a) On the revenue side, maximum contribution is done by timber and resin, followed by Tendu leaves, fuel and bamboo.

(b) The revenue from timber increased from Rs.2,502.95 lac in 1975-76 to Rs.3,147.80 lac in 1978-79. It declined in 1979-80 to Rs.3,064.63 lac and since then it has been increasing and has reached to Rs.4,015.45 lac in 1981-82. Similarly, in fuel the revenue increased from Rs.87.07 lac in 1975-76 to Rs.142.93 lac in 1976-77 and then declined to Rs.99.12 lac in 1978-79. Since then it is increasing with the minor decline in 1980-81 and has become Rs.133.42 lac in 1981-82.

(c) In the case of bamboo the revenue declined from Rs.35.86 lac in 1975-76 to Rs.15.25 lac in 1977-78 and since then it has been constants rising and has reached to Rs.57.58 lac in 1981-82.

(d) The revenue from Resir declined from Rs.413.44 lac in 1975-76 to Rs.244.21 lac in 1976-77. It is since 1976-77, that the revenue from resin has constantly increased and reached to Rs.639.47 lac in 1980-81. It declined in 1981-82 to Rs.578.37 lac.

(e) In the case of revenue from tendu leaves, there were very frequent fluctuations, with increase and decrease taking place every alternate year. The revenue from tendu leaves increased relatively less as compared to other sources. It increased from Rs.221.88 lac in 1975-76 to Rs.282.15 lac in 1981-82.

(f) The total revenue constantly increased. It increased from Rs.3,685.30 lac in 1975-76 to Rs.5,747.62 lac in 1981-82.

On the expenditure side, maximum expenditure took place in field i.e. work and on administration. The expenditure on work constantly increased. It increased from Rs.970.67 lac in 1975-76 to Rs.2,346.79 lac in 1981-82. The expenditure on administration also increased from Rs.518.54 lac in 1975-76 to Rs.1,032.07 lac in 1981-82. The miscellaneous expenditure increased gradually from Rs.36.17 lacs in 1975-76 to Rs.43.54 lac in 1977-78 and then declined to Rs.42.95 lac in 1978-79. But in 1979-80 the miscellaneous expenditure suddenly increased and more than double itself. It became Rs. 98.39 lac in 1979-80 and since then it is rising and has reached to Rs.255.04 lac in 1981-82.

The total expenditure has constantly increased from 1975-76 and has become double in 1981-82. The total expenditure increased from Rs. 1525.38 lac in 1975-76 to Rs.3633.90 lac in 1981-82.

The net revenue from forests showed a fluctuation. The net revenue increased from Rs.2159.92 lac in 1975-76 to Rs.2262.43 lac in 1976-77 and declined to Rs. 2113.90 lac in 1977-78. In 1978-79, the revenue increased to Rs.2176.33 lac and then again declined in next two years ~~and became~~ Rs.1748.16 lac in 1980-81. In 1981-82, the revenue again increased to Rs.2113.72 lac. The fluctuations in net revenue took place because of the increasing expenditure of the forest department. The increase in expenditure took place because of increase in employment done for plantation and afforestation purposes and because of social forestry programme.

In the table 11 one thing is of great importance that total revenue is less than the total value of the forest product for the same year. After enquiry from the Statistical Department it was revealed that word Value means the cost of the product, in actual, while revenue stands for the cost of the product at the production stage. In value, all taxes, profits, transportation and other processing costs are included. This explanation is not correct.

The method of extraction of forest produce in Uttar Pradesh was by the contractor system only. In 1966, the Government (Department) started the work in certain areas of the State, to check the illegal practices adopted by the contractors. In 1974, the Forest Corporation--an organization within the department came into being and gradually abondoned the former agency of exploitation, and also undertook the entire control and management of processing and marketing of the forest products.

The total out turn of timber extracted by the Forest Corporation, U.P. during the last 10 years is given in table 12:

The table depicts that in the initial year of formation of U.P. Forest Corporation i.e. 1974-75, the Corporation only extracted Sal and little of Chir and Deodar. But from the very second year the total produce extracted by the Corporation increased to 92,000 cubic meters of which 47,000 cubic meters was Sal, 29,000 cubic meters was Chir and 16,000 cubic meters was Deodar. In 1976-77, the Corporation extracted

Table-12

	In Thousand Cubic Meters					
Year	Sal	Khair	Chir	Deodar (Fir,Kail)	Eucalyptus.	Total
1974-75	23	-	1	1	-	25
1975-76	47	-	29	16	-	92
1976-77	68	13	51	-	-	132
1977-78	63	10	37	21	-	131
1978-79	60	20	120	30	-	230
1979-80	72	13	123	65	-	273
1980-81	77	17	169	25	17	305
1981-82	114	10	246	23	86	479

SOURCE : Forest Statistics, U.P., 1982, p.54, U.P. Forest Deptt.

1,32,000 cubic meters of forest produce, which included Sal, Chir and Deodar. From 1980-81, the Corporation also undertook the extraction of Eucalyptus, thereby increasing the total extraction to 3,05,000 cubic meters. Since then the Corporation has never looked back and its activities are constantly developing.

I. (d) MAIN INDUSTRIES

By virtue of possessing both temperate and tropical forests, Uttar Pradesh is endowed with a vast variety of natural forest vegetation which yields many kinds of timber, both hardwood and softwood, bamboos, grasses, herbs, medicinal plants, canes and a rich fauna. These, in turn offer a bewildering variety of minor forest products. Many an industry both large and cottage scale, thrive on the products of the forests. The principal industries of Uttar Pradesh are mentioned below:

MATCH INDUSTRY

Prior to the first World War, nearly all the requirements of the matches in India were met by import mainly from Sweden and Japan. From 1924, a large scale mechanised match industry developed. However, the only match factory in this State was started in 1929 by the Western India Match Co.Ltd. at Callutterbuckganj, Bareilly. This factory employs nearly 2000 workers and roughly produces about one lakh fifty thousand (1,50,00) match cases, each case containing fifty (50) gross match boxes, valued at approximately Rs.two crores per annum. The total requirement of this factory is about six lac cubic feet of match wood per year, out of which 30 per cent of the timber is supplied by the U.P. Forest Department from Tarai and Bhabhar area and some from Nepal. The timber supplied are mainly semal (Salmalia malabaica) and gutel (Toewia nudiflora).

As the industry is of great importance, an attempt is being made under the five year plans to increase the plantation of matchwood species like ; Semal, Aru (Ailanthus excelsa), Gutel, Alstonia scholaris, etc It is expected that these plantations on their maturity will make the establishment of as many as nine factories of 20,000 tons capacity possible by 2016 A.D. Presently, there are only four match industry in the State.

Number of trees allotted to match industry by the Forest Department in 1978-79 and 1979-80 were 9,916 and 5,652 respectively. In 1980-81 only 4,137 trees were allotted[16].

PLYWOOD

There are at present three Plywood factories in Uttar Pradesh. The biggest one is situated in Sitapur, known as Sitapur Plywood products. It was started in 1940, after the Burma Campaign for the manufacture of Plywood. The other two are situated at Jwalapur and Najibabad. The present annual production is about 20 lac sq.ft. which is less than the rated capacity. Only part of the requirement is met by U.P. Forest Department and the balance is procured mostly from Madhya Pradesh and Assam. It is estimated by 2016 A.D., there is possibility of establishing five new units for Plywood manufacture each with a capacity of 12 m.sq. ft. of commercial Plywood of 3 mm and 4 mm thickness.

The waste of the Plywood factory is utilized in making the particle board. There is a particle board plant in Sitapur with production capacity of 10,000 sq.ft. of particle board per day. On account of limited market, the present production is very less. Mostly, the product is used in the manufacture of veneer board, doors, etc. in the Plywood factory itself. Number of trees allotted to Plywood factory in 1978-79 and 1979-80 were 15,394 and 14,120 respectively. In 1980-81, the number of trees allotted were only 11,182.[17]

PAPER AND PULP INDUSTRY

The supply of paper is inadequate, at reasonable rates, for meeting the multifarious needs for paper in a civilized society. The present per capita consumption of paper in the State is about 1 kg which is likely to go upto 4-5 kg in about 20 years.

At present, there are only two paper mills in the State. The Star Paper Mill at Saharanpur and the Upper India Paper Mill at Lucknow. The raw material used in the former is Sabai or Bhabar grass (Eulaliopsis binat) and in the later Sabai grass as well as waste paper cuttings and rags. The production of both the mills was 10,000 tons, but with the recent expansion of the Star Paper Mill, the production has gone upto 24,000 tons approximately.

Besides this, there are cottage scale industries making hand made paper in Kalpi (Kalaun) and Kalsi (Dehradun). In Kalpi, more than 400 families are engaged in the production of paper. They use waste paper and rag as their raw material.

The total amount of Baib grass used by the Star Paper Mills and other industries is nearly 59,142 quintals which amounts to Rs.11,14,940/- in 1978-79. The Star Paper Mill used twisted Chir amounting to 12,393.50 metric tonnes. Since 1979, the U.P. Forest Department has been able to meet the increasing requirement of twisted chir. This is clear from the table given below:

Twisted Chir supplied to Star Paper Mill:

1979-80 - 12,991.04 metric tonnes.

1980-81 - 12,835.71 metric tonnes.

1981-82 - 20,706.43 metric tonnes.

SOURCE : Own compilation on the basis of Forest Statistics, U.P. Forest Department.

KATHA AND KUTCH INDUSTRY

Katha, the major ingradient of Pan is extracted from the heartwood of Acacia catechu, which yields5-6 per cent of Katha and in addition 10 per cent Kutch (Cutch) which is at present separately recovered only to a small extent.

There are eight Katha Factories in U.P. located at Izzatnagar, Bareilly, Haldwani and Najibabad. The principal factory of Izzatnagar is known as the Indian Wood Product Company. It is in existence for past 50 years or so, while the others are of present origin. The factory at Izzatnagar processes about 10,000 tons of Katha wood and produces about 500 tons of Katha and 1,000 tons of Cutch. The remaining factories utilize about15,000 tons of heartwood and produce about 400 tons of Katha and 1,000 tons of cutch. Their annual capacity varies from 1,000 to 3,000 tons of heartwood.

The Khair chips after the extraction of tansin are at present used in the Katha furnaces, while they can be used for the manufacture of particle board. Large-scale plantation of Khair have been raised during the 5 year plans. It is expected that he potential yield of 69,000 tons of Khair heartwood by 2016 A.D. will yield about 3500 tons of Katha by the factory process and sustain about 6-7 additional unit, each consuming about 10,000 tonnes of heartwood per annum.

Cutch after blending with other extracts (Babul, Myrobolans, Wattle, etc.) is used as a tanning material for curing a few types of leather, but its main use is in a colouring dye. It is now increasingly used in well drilling operations by the O.N. G.C. (Oil & Natural Gas Commission).

Besides this, Katha is produced by indigenous methods of chair in most of the cottage industries. Katha is produced in this manner in Nainital, Gonda, Lakhimpur-Kheri, Sharanpur, Bijnor, Bahraich, Garhwal and Dehradun districts. The quality of Katha produced is slightly inferior.

SPORTS GOODS INDUSTRY

The sports goods industry is a post partition enterprise in the state and is located at Dehradun Meerut, Agra, Allahabad, Lucknow, Kanpur and Nainital. The main articles manufactured are :Hockey sticks, Cricket bats, Tennis and Badminton Rackets. A scheme for the development of sports goods industry has been started at Bareilly by the industries department with the intention of improving the product and also training 20 persons every year. Meerut is the biggest centre of the industry, having 87 units and employing nearly 1500 persons and manufacturing sports goods worth Rs.32 lacs. Forest Department allotted nearly 1873 trees during 1979-80 and 1,263 trees during 1980 -81. The forest species supplied to sports goods industry are Haldu, Gutel Mulberry, Faldu, Pula and Kanju.

RESIN INDUSTRY

One of the most important forest based industry in the State is of resin, which is derived from Chir pine(Pinus roxburghii). Resin is a source of resin and turpentine, which are largely used in the manufacture of soap, paints, varnishes, paper and disinfectants. Turpentinealso provides ample scope for a profitable use in the manufacture of menthol and synthetic camphor.

Crude resin is obtained from the stem of chir pine trees which occur abundantly in the Kumaon and Tehri circles of the State. The extraction is done departmentally through mate system. A major part of government tapped resin is allotted to the Indian Terpentine & Resin Co. Limited, Bareilly, 80 per cent share of which are held by U.P. Government. This company was started in 1919 and produces about two lac mds of resin and two lac mds of turpentine.

The quantity of resin extracted from the U.P. Forests under normal tapping programme is of the order of 1.5 lacs quintals. The total yield in 1978-79 was about 1,29,609 quintals, costing nearly 3 crores of rupees (Rs.3,70,00 549). The crude resin extracted is supplied to the I.T.R. Cooperative Societies, Small Scale Industries and sale by auction. The total supply then was about 95,000 Quintals to I.T.R. Co., 11,000 Qtls. to Cooperative Society, 17,000 Qtls to small scale industries and rest of by auction. The total supply in 1981-82 was 86,065 Qtls to I.T.R. Co., 5,437 Qtls to Cooperative Societies, 35,475 Qtls to Small-Scale Industries and 11,310 Qtls by auction, thus amounting to 1,38,287 Qtls. The total supply of resin from 1979-80 to 1981-82 is shown as follows:

In addition to I.T.R. Co., Bareilly, there are cooperative factories in Rishikesh, Kotdwara and Ranikhet. There are two factories in Haldwani, one in Bareilly, one in Lucknow which depends on the resin tapped by privately owned lands.

BIRI INDUSTRY

The leaves of tendu (Diospyros to mentosa), are used for the manufacture of Biri, the Indian cigarette. Allahabad, Moradabad, Varanasi, Jaunpur, Farrukhabad, Jhansi, Azamgarh, Mirzapur, Banda, Ghazipur and Ballia are the chief centres of Biri industry. On the average, about 37,000 persons are employed in this industry. The capital investment is estimated at over Rs.75 lacs.

RESIN SUPPLIED TO I.T.R. CO., BAREILLY, CO-OPERATIVE SOCIETIES, ETC. (IN QUINTALS)

Sector / Year	*1979-80*	*1980-81*	*1981-82*
I.T.R. Co., Bareilly.	91,993	1,25,784	86,065
Co-operative Societies.	11.113	6,875	5,437
Small-ScaleIndustries.	29,187	37,206	35,475
By Auction	848	4,751	11,310

SOURCE : Own compilation on the basis of Forest Statistics, U.P. Forest Department.

FURNITURE MAKING, CARVING, ETC.

Workshop for making furniture exists in almost all the towns, big and small. Wooden agricultural implements and tool handles are made everywhere by carpenters. Industries Department runs mobile carpentry units with a view to teach village carpenters improved scientific methods of wood working. The main centre for wood caring is Saharanpur. Carved and ornamental cigarette cases, trays, small screens, wall brackets, tables, book racks, etc. are made of Shisham (Dalbergia Sissoo) wood. Wooden toys are made at several places in the State. The important centres for the production of wooden toys are Varanasi and Amroha (Moradabad) where Kurchi (Holarrhene antidysentrica) and dudhi (Wrightia tomentosa) woods are used.

TANNING INDUSTRY

Kanpur and Agra are important centres of leather trade in the State. Primary tanning material used is babul and wattle of which babul is produced locally and the latter is either fully imported or partly imported from Madras. Among the secondary materials, banks of Cassia auriculata and C. fistula in addition to fruits of myrobalan (Terminalia chebula) are commonly used.

ESSENTIAL OIL INDUSTRY

Uttar Pradesh is rich in perfume bearing plants and grasses and has a well developed essential oil industry of which Kannauj is an important centre. Distillation of Vetiver oil from has (Vetiveria Zizanioides) is done. It is the most important industry. About 2000 of vetiver oil is produced in the State. Rosha oil is produced from Cymbopogon martinii at Bibiwala in Dehradun district

BAMBOO, RINGAL AND CANE

The U.P. State forests produce a lot of bamboo, ringals and canes, which are used in the manufacture of furniture, baskets, boxes and other things of daily use, like mats, fans, etc. Besides this, grasses and climbers is used for the manufacture of ropes and strings.

Besides these main industries, there are many other industries like packaging industries, pencil, battery separator industry, medicinal herbs industry, etc. There is also large supply of sleepers to railways amounting to 35,090 cu.m., costing Rs.2,79,94,787/- in 1978-79 and 2,553 cu.m.of timber to D.G.S. &D. which amounted to Rs.18,79,733/- in 1978-79. The total timber supplied to Railways and D.G.S. & D. between 1979-80 and 1980-81 is given in the table below:

The above table shows that the supply by the Forest Department to the Railways and D.G.S. & D. constantly increased and amounted to Rs. 42,32,350/- in 1980-81.

Sector	YEARS			
	1979-80		1980-81	
	Quantity (Cu.m.)	Value (Rs.)	Quantity Cu.m.)	Value (Rs.)
Railways	25,902	2,97,43,471/-	35,609	4,39,57,941/-
D.G.S. & D.	255	3,37,565/-	259	2,74,409/-

SOURCE : Own compilation on the basis of Forest Statistics U.P., U.P. Forest Department.

There are new other minor products like gum, gypsum, lime and lime stone, honey, wax, hides and horns, thatching and other grasses.

The forests of the State of U.P. are not in accordance to the scientific criterion of 33.33% of the total geographical area. Besides this, the existing forests are also not properly distributed between the hills and the plains, thus causing ecological imbalance. The distribution of forests between hills and plains is 48 per cent and 7 per cent respectively, whereas it should be 60 per cent and 40 per cent respectively. This imbalance makes proper exploitation and utilization necessary and also gives emphasis to proper development of forests. Lastly, the forest of U.P. are having large industrial potential and are of high economic utility. A proper exploitation and utilization of forests can further help in teh development of forest based industries which will in turn increase the level of employment. To improve exploitation and utilization, it is necessary to remove the defects of processing and marketing of forest products. Removal of such defects will help not only in increasing the revenue to the State Exchequer but also in generation of resources, which will prove beneficial to the economic development.

References

1. G.N. Singh, "Forest Administration and Public Opinion, p.42. The Centenary of Forest Administration in Uttar Pradesh, 1861-1961, U.P. Forest Department, 1961.
2. G.N. Singh, "Forest Administration and Public Opinion, p.43. The Centenary of Forest Administration in Uttar Pradesh, 1861-1961, U.P. Forest Department, 1961.
3. A Brief History of Forest Administration in Uttar Pradesh, p. 19, The Centenary of Forest Administration, 1861-1961, U.P. Forest Department, 1961.
4. G.N. Singh, "Forest Administration and Public Opinion, pp.42-43, The Centenary of Forest Administration in Uttar Pradesh, 1861-1961, U.P. Forest Department, 1961.
5. S.P. Sahi, "Forests-Their Geographic & Economic Influences." pp 83 - 84, The Centenary of Forest Administration in Uttar Pradesh, 1861 - 1961, U.P. Forest Department, 1961.
6. S.P. Sahi, "Forests - Their Geographic & Economic Influences" pp. 83 - 84, The Centenary of Forest Administration in Uttar Pradesh, 1861-1961, U.P. Forest Department, 1961.
7. S.P. Sahi, "Forests - Their Geographic & Economic Influences" p. 84, The Centenary of Forest Administration, 1861-1961, U.P. Forest Department, 1961.
8. S.P. Sahi, "Forests -- Their Geographic and Economic Influences" p.85, The Centenary of Forest Administration in Uttar Pradesh, 1861 - 1961.
9. (For details refer : Appendix - I).
10. T. Mehta, A Handbook of Forest Utilization, pp. 123 - 155, Periodical Expert Book Agency, Delhi.
11. For details of medicinal herbs, refer - Appendix - II.
12. R.C. Soni, Forestry of Uttar Pradesh, Hundred Years of Forestry in U.P., 1861 - 1961, U.P. Forest Department, Lucknow.
13. G.N. Singh, "Forest Administration & Public Opinion", p. 44, The Centenary of Forest Administration, 1861 - 1961, U.P. Forest Department, 1961, Lucknow.
14. A Brief History of Forest Administration in U.P., p. 21, The Centenary of Forest Administration in U.P., 1961. U.P. Forest Department.
15. R.C. Soni, 'Forestry of Uttar Pradesh', pp. 24 - 25, The Centenary of Forest Administration in U.P., 1861 - 1961, U.P. Forest Department, 1961.
    Champion, H.G. and Seth, S.K. - 'Forests Types of India', Government of India.
16. Forest Statistics, U.P., U.P. Forest Department, Lucknow.
17. Forest Statistics, U.P. Forest Department.

CHAPTER TWO

Former Pattern of Working - Contractor System, Purpose And Evaluation

Forest is one of the most vulnerable natural resources of the country. Forest existed in perennial plenty when man's need was small. Forests were at the door step of man in the past, but as he emerged from the dark ages and progressed through a vast space of time down to the age of industrial revolution, the vast tracts of virgin forests saw wanton destruction at the hands of man in order to meet the increasing demand of rising population and for the establishment of new factories, work-shops, railways, etc. The unscientific working shrank the forest and made it scarce. In certain places/areas afforestation also became economically infeasible. This gave birth to an urgent need for managing forests scientifically and also utilizing them judiciously, so that actual production of it may increase.

The scientific method of conservation, based on the silviculture system, has evolved through a long history of management. During the ancient period, Katyuri (500 B.C. to 700 A.D.), Chand (700 - 1790 A.D.) and Parmar Kingdoms, forests were managed by 'Manusmriti'- the ancient scripture on law, which holds Vapi (Water Point), Koop (Wells), Tarag (Pond), Devalaya (Temple) and trees as public property.[1] These were managed by the 'King', as the head of the State. During this period there was no restriction or check on the exploitation of forest resources. But by the end of the seventeenth century, the encroachment on forest area started by Rohillas in the area east of Ganges i.e. Tarai Bhabar and Sikhs in the Dehradun valley. These incessant raids affected the control on forests and resulted in the damage of forest area.

It was in the 1790 and 1805, the fall of Kumaon and Garhwal took place at the hands of Gorkhas. With the fall the entire forest area became the property of Gorkhas. [2] Under the occupancy of Gorkhas, massive destruction of forest took place. The Gorkhas destroyed these forests to protect their fortresses from the rebels and local outlaws as they took shelter in the forests. The Gorkhas, for the first time started collecting tax on grazing in the shape of 'Gheekar' or ghee and levied taxes on timber, Bamboo and Kutch (Catechu). It was in the year 1815, the Britishers defeated the Gorkhas and took over the administration of the forest area. They took a long time to provide stable administration to the troubled land.

In the initial years of British occupancy dues styled as Kathbans (Tax on Timber & Bamboo), Kathmahal (Tax on Katha), and Ghikar (Grazing tax as Ghee or clarified butter) were only collected.[3] But later on these forest areas were leased to the local Zamindars of the parganas and dues in the form of excise duty and transit duty on merchandise were collected at the posts erected at the passes. From 1819-21, Major Young, offered the right of levying the forest duties on the public auctions.[4] In 1823, Mr. Trail, during land settlement operations demarcated the village boundaries for grazing, cutting timber and collecting firewood. In 1825, Wallich was deputed by the Government to examine the extensive forest area at the foot of Himalayas. He recommended the conservation of the forests of Sal and Shisham in Avadh and Tarai regions. But the department never felt the necessity of checking even undue exploitation of forests - thinking that the forest wealth of India is inexhaustible. The unchecked exploitation of accessible forests, the forest fires and grazing added to the cause of gradual destruction of forests in the plains of Uttar Pradesh.[5] In 1826, Mr. Trail prohibited the cutting of Sal and the other forest produce from all the 'thaplas' throughout Bhabar.[6] But in the Dehradun Valley, Sal forest was heavily overworked causing the accessible portions to be depleted of valuable stock.

In 1840, Mr. Wilson obtained leases for exploitation of forest produce on nominal considerations. Magnificient large size trees were felled and the converted timber was floated down the Bhagirathi in high floods. Salvage of even a fraction of this timber was economical to Wilson since he had obtained it almost free of cost. The devastating effects of over-exploitation is visible to this day. According to Mr. William, Commissioner of Meerut, the State of forest was in bad shape due to over exploitation. "Every one continued to hack and how away at the trees as he pleased, only paying certain dues to the former, in the event of wood being exported."[7]

Recklessness was inevitable and the fine Sal forests began to disappear rapidly, because of absence of conservancy. Till 1855, the destruction of forest that took place was incalculable. In the period between 1855 to 1861, an added impetus was given to the uncontrolled exploitation by the ever increasing demand for railway sleepers, charcoal for running railways and timber for their navy.[8] To carry out this work, the British contractors were given uncontrolled liberty to cut and fell where they pleased and whatever way they pleased. With the result they felled more timber which outstripped their capacity to transport and export. This resulted in accumulation of timber as debris. It took forest officials several years to extract the timber thus left by contractors.

The commencement of a new era for forestry in India is marked by the year 1855, as it was in this year that Lord Dalhousie, the then Governor General of India, laid down the future policy in a memorable announcement. From 1860 onwards forest administration was extended and developed in various provinces. In Uttar Pradesh, then known as North West Provinces, forest administration started from the year 1861. In 1861, Colonel Sir Henry Ramsay, the then Commissioner of Kumaon and Garhwal was appointed as ex-officio Conservator of Forests.[9] Col. Ramsay was quick to recognize the gravity of the situation and he took immediate measures to stop this wanton destruction of forests. To check further destruction of forests and to provide place to this valuable property on a sound and more prominent footing, he took the following initiatives to introduce rational exploitation :

(a) Abolished shifting cultivation and leases for farming and started systematic management.

(b) Indiscriminate felling of trees was stopped and a system of conservancy began under which hammer-marking before felling of trees was arranged.

(c) Introduction of Forest working on rotational basis.

(d) Grazing was regulated and cultivation of land in valuable Sal forest was prohibited. Farmers were given land below submontane road in the less valuable forest.

(e) Protection of fire in forest was undertaken.

The enforcement of control resulted in diversion of forests from private speculators pockets to the state treasury. In 1864, Dr. Dietrich Brandis was appointed as the first Inspector General of Forests.[10] It was the result of his initiative and drive that regular forest department came into being in 1868, in Uttar Pradesh. By 1877, most of the valuable forests in sub-himalayas, Taras-Bhabar containing Khair-Sissoo was demarcated and mapped.[11] During this period, green timber was worked under heavy thinning made by departmental agency and the contractors were only allowed to remove the fallen wood and dead wood left on the ground from the past felling. Work continued ahead and roads buildings construction fire protection was pushed on. Lease of minor forest produce and dead wood was replaced by the Kham-Tehsil System (By which payment for produce was made through forest 'Chaukis' on export from the forest). This system was also abolished in 1881.[12]

As a first step toward regulation and control, the Government of India passed the Act VII of 1865 for the management of State Government Forests. This Act was later on replaced by an elaborate Forest Act XVI of 1878, which provided for the constitution of 'Reserved' and 'Protected' forests.[13] In 1893, all the unmeasured forests were declared protected under a proclamation. It also declared certain trees as reserved trees which could not be cut without a permit. Various restrictions on extension of cultivation, felling, lopping of trees, shooting, hunting and fishing were imposed. This proclamation did not pay any heed to the rights and concessions enjoyed by local population. The proclamation was replaced in 1903 and the protected Civil Forests were divided into : (A) Closed Civil Forests and (B) Open Civil Forests. In 1911, a new step was taken which categorised forests in the following heads :

(A) Reserved Forests (i) Old Reserves.

(ii) New Reserves.

(B) Protected Forests or Civil Forests.[14]

As a result of this settlement, curtailment of the rights and concessions of villagers took place which resulted in resistance, violence, incendiarism and illicit felling.

The Reserved Forests were under the control of Forest Department while the Civil Forests were managed by the revenue department. Due to maltreatment by the revenue department, the class I forests suffered heavily, as there existed no restriction on grazing, felling and looping of trees. A stage reached when at some places the class I forests reached their extinction, affecting the dependent village economy adversely.

The Forest Act XVI of 1878 was later replaced by the Indian Forest Act, XVI of 1927.[15] Various circles were created during the Second World War to meet the demand of timber for war purposes. After the war, when India gained independence, the royal forests of Tehri State were merged with State Forest and a Tehri Circle was created. Some Zamindari Forests were also merged with the state forests. Though, some of the forests which were merged were in good condition, yet many of the forests had depleted stocks and needed proper rehabilitation. Later, various schemes were implemented under the period of planning. To cope with the increased activities of forestry, various scientific methods of management like silviculture systems, working plans, etc. were introduced.

Since its inception, state forest department took over those areas which were accessible but more or less in ruined condition. They were full of overmature, hollow or badly shaped trees and were also subjected to indiscriminate burning and grazing. When these degraded forests having depleted stock were transferred to the forest department, as a first step, the department planned to get rid of the unsound trees wherever possible, provided these were no longer required as seed bearers. The surplus old growing stock was exploited and utilised for railway sleepers. Cutting of climbers and removal of weeds was also done.

During the period 1948 to 1960, the following Acts were passed [16]:

1) The U.P. Private Forest Act, 1948.

ii) The Zamindari Abolition and Land Reforms Acts, 1950.

iii) The Indian Forest (U.P. Amendment) Act, 1960.

The passing of these Acts caused the area under the control of Forest Department to increase many fold. For several decades the forest department remained engaged more in the conservation of resources than their exploitation. The department introduced scientific principle of forestry management and development. These principles were based on the working plans prepared for each division for a specific period of time and which were subjected to revision after regular intervals. These working plans gave guidelines for the proper management, development that is plantation and afforestation and exploitation of the forest areas.

The exploitation of forest resources from the state forests was an important task of the forest department administration. The U.P Forest Department managed the forest resources on the basis of silviculture system of management and the working plan prescriptions. The department decides the various aspects of exploitation of forest resources and also regulates the amount of exploitation to be done through various agencies.

The extraction or marketing or forest produce was done through the following agencies, namely :

i) By Government agency or departmentally.

ii) The Government and the purchaser.

iii) By purchaser alone.

It was very difficult to say which system of exploitation is the best and most remunerative for any given set of circumstances. Now to select one of them, the following considerations were to be borne in mind :

i) The maintenance and improvement of forest,

ii) A fair return to the government,

iii) Preventing theft and fraud by the forest employees and purchasers,

iv) Physical conditions of the locality,

v) The interest of the local population,

vi) The availability of staff,

vii) The availability of skilled labour,

viii) The availability of labour, and

ix) The avoidance of unnecessary complication.

The above factors differ from one locality to another and thereby make the choice of the agency evident. The different systems of exploitation are discussed in the following paras :

Department /Government Agency

Departmental system of working was introduced by Dietrich Brandis in 1864 in the sub-Himalayan Forests.[17] Under this system the Government department had to carry out the entire work of felling of trees, conversion, extraction i.e. transportation and disposal of all types of products. The departmental system prevailed only for two years and then it was changed, as it proved uneconomical from the view point of revenue. Later on, the felling and extraction of timber to the sale depot was done by the persons/ contractors working on behalf of the government. The contractors were paid according to the contract between them and the Government. The government then organised the sale of the product from the sale depots. The net profit earned by the government being the difference between the sale value and the working expenses. Subsequently, due to various limitations, the departmental system of extraction was only restricted to meet the departmental needs and commitments. In order to do so, the forest department hired labourers on daily wages and got the extraction of timber done under the supervision of its staff. The produce thus, extracted, was brought to forest godowns from where it was sent to desired places. Till recently, the extraction of minor forest produce i.e. resin, baib grass, etc. was done through departmental agency.

The departmental system of exploitation was not much profitable owing to the following reasons :

a) The work was mostly done by hired labour on daily wages (Sirolis and Nagpuris), who were not careful about their efficiency.[18]

b) The supervising staff was not strict. They did not take keen interest in getting the fuller amount of produce extracted.

c) The supervising staff was not trained in logging operation. He often went to the spot of work only to show his presence and not the aptitude as shown by the contractor.

d) Lastly, red tapism also hinders in the efficiency of worker.

The Government usually does not take to departmental extraction but prefers private enterprise. It only resorts to departmental agency when it has to meet the regular supply of firewood or charcoal, or to keep down the prices of forest products, or when the work can not be given to the contraction because of its tedious nature.

The Government and Purchaser Combined

This system is adopted to exercise control over a part of the work undertaken. For example, if the department is fearing big loss of forest during felling then in such case the departmental agency does the felling work while the extraction and selling is done by the contractors. This system has proved beneficial and suitable for the places where thinnings or improvement fellings are to be carried out.

Purchaser Agency

The purchaser system or the contractor system is the most common system of working. It is estimated that the major share of production of wood, bamboo, etc. is exploited by contractors. Under this system the felling, extraction and collection of forest produce is solely done by the contractor. In this system, the forest staff is free from the burden of additional work. This system also encourages healthy competition in the private enterprise.

In a big state like Uttar Pradesh, much importance was given to the contractor system owing to the following reasons :

1) The forest department has to do nothing except stamping (marking) the trees.

2) Forest staff is free from the burden of unnecessary work.

3) The risk of the department is shifted over to the contractor.

4) It helps in increasing the revenue of the department, by competitive bidding in open auctions.

In the contractors system of working, the forest department gives the contract of a forest coupe (lot) or a number of coupes by auction to the purchasing contractor on certain terms and conditions, for the purpose of exploiting forest produce. The contractors then make arrangement to fell the trees, to saw it and to extract it from the forest area to its depots through various means. The contractor engages its own labour and supervises the work either himself or gets it supervised by his kith and kins. Sometimes big contractors employ sub-contractors to harness labourer from nearby areas. These subcontactors arrange the supply of rations and technical hands like Munshis, Moharrirs, etc. After completing felling of produce it is taken to the depots (Contractors), which are in or near the marketing centres and mandis for the purpose of sale and further export.

The total out turn of major forest produce as produced by Government and Purchasers system during 1978-79 to 1981-82 can be studied from the following table :

i) The out turn of timber by departmental increased from 8,707 cu.m. in 1978-79 to 9,519 cu.m. in 1979-80, then decreased to 8,987 cu.m. in 1980-81. The out turn of timber increased in 1981-82 to 13,867 cu.m.

Table - 13

OUT TURN OF MAJOR FOREST PRODUCE BY DEPARTMENT & PURCHASER.

TIMBER-	*IN ROUND*			*IN Cu. METERS*
Agencies	*1978-79*	*1979-80*	*1980-81*	*1981-82*
Departmental	8,707	9,519	8,987	13,867
Purchaser	10,82,860	8,81,445	7,37,998	8,78,750
Total	10,91,567	8,90,964	7,46,985	8,92,617
FUEL				*IN Cu. METERS.*
Agencies	*1978-79*	*1979-80*	*1980-81*	*1981-82*
Departmental	92	195	582	6,149
Purchaser	15,98,400	16,13,492	10,99,525	10,74,942
Total	15,98,492	16,13,687	11,00,107	10,81,091

SOURCE : Forest Statistics, U.P. Forest Department, p. 50 (1981-82).

ii) The out turn of timber by purchasers system decreased from 10,82,860 cu.m.in 1978-79 to 7,37,998 cu.m. in 1980-81 and then again increased in 1981-82 to 8,78,750 cu.m.

iii) The fuel production by departmental system constantly increased from 92 cu.m. in 1978-79 to 195 cu.m. in 1979-80, 582 cu.m. in 1980-81 and 6,149 cu.m. in 1981-82.

iv) The out turn of fuel by purchaser initially increased from 15,98,400 cu.m. in 1978-79 to 16,13,492 cu.m. in 1979-80 and then decreased to 10,99,525 cu.m. and 10,74,942 cu.m. in 1980-81 and 1981-82 respectively.

The fall in out turn of timber and fuel by the purchasers and increase in out turn by departmental system is because of the shift in production from purchasers system to departmental working, due to the emergence of forest corporation.

System of Sale

There are various methods by which the forest department may sell timber and other forest produce. These systems can be grouped under the following heads.

1) Lump sum sale

2) Payment on out turn.

Lump sum sale

Lump sum sale ensures a fixed sum of money to the forest department for a lot of logs, an area of grass, a fuel coupe, a lot of trees marked for felling, etc., of which the exact quantity of produce is not known. In this system, sale is done by auction, tenders or directly. After the sale the purchaser is licenced or permitted to extract the produce from a certain specified area of forest for a fixed sum. The exact amount of produce is only known when it is extracted. This also covers lease or a contract for extraction of marked trees. If the purchase price is paid on lump sum basis then the price is realised on instalments, with the first instalment being paid before starting the actual extraction.

This (Lump sum sale) form of sale is best, if the amount of produce saleable is known, for example, fuel or minor forest produce at a sale depot. But as regards the standing trees and unknown produce it is objectionable as the amount proportional to the produce extracted can not be known accurately.

Payment on out turn

All types of sale come under payment on out turn as in the produce extracted corresponds with the price paid. This system has produce extracted corresponds with the price paid. This system has many varieties of extraction and sale. The following varieties were practiced in Uttar Pradesh Forests.

(a) Sale of a whole coupe or area

In this system a certain area or portion of forest is given on lease for a specified period of time to a person or firm, to extract timber and other forest produce. The system of payment is lump sum in one premium or in instalment, or royalty basis or sometimes by a combination of both. This system is not suitable for the extraction of timber or fuel wood, as this would lead to over-exploitation of area. But in the case of minor forest produce like gum, honey, grass, lac, etc., it is the best. On the other hand it is the worst method in case of dead timber because the purchaser will be induced to cut the living stock, if guided by self interest. The time of lease also varies with the prevailing circumstances.

(b) Sale of standing tree

The sale of marked standing tree is the best suited to the forest department. In this system the trees are first selected and marked and then the whole lot is sold standing by either auction, tender or private bargain directly. Sale may be either by lump sum payment for the coupe or by royalty on out turn based on rates per tree or on per cubic foot (or metre).

A well known variety of this system is the monopoly system in which the trees marked in a coupe are sold by auction, the price paid is the monopoly price which gives the purchaser the right to fell and extract timber. Besides this monopoly price, the purchaser pays the royalty on the extracted timber. This system has certain drawbacks, like, if the royalty rates are high then the purchaser is not induced to extract inferior timber from remote in accessible areas. While on the other hand, if the royalty is low, the purchaser will be induced to extract everything, as he will be earning more profit. The fixation of royalty is not an easy job, that is why, the lump sum sale is preferred.

(c) Sale by licence and permits

In this system the intending purchaser has to apply for a licence to extract timber or other forest produce from the notified forest area in a given time. A fee fixed by the forest department is paid by the Licencee for the licence, after which he is permitted to collect the produce. This method is generally used to collect the minor forest produce.

(d) Sale of few selected species

In this the sale is usually done by private bargain or by fixed tariff of a few selected trees, in special cases of sale to public departments, institutions, sporadic demand of local people, and also the sale of certain species which are demanded for particular purpose or industry.

Method of Sale

Thus there are following common methods of sale of forest produce:

1) Sale by private bargain (treaty).

2) Sale by fixed tariff.

3) Sale by auction.

4) Sale by tenders.

(1) By private bargain

This is the most ordinary method and is used in some special circumstances. For example, the produce not being of general demand but is demanded by a particular person or firm. In such cases, the department deals with the single purchaser and negotiates the price after mutual discussion. In private bargaining, the prices are kept at lower level so as to promote the interest of other buyers. This method is used under following circumstances :

a) When the quantity of produce to be sold is small.

b) When the produce is having limited demand, i.e. when it is demanded by a particular person, firm or an industry.

c) To popularise the use of a particular species for a new industry.

d) To break the auction rings.

(2) By fixed tariff

This method is similar to private bargaining except that the prices are fixed before hand and can not be changed without official sanction.

The prices generally resemble the current market rates and are subjected to revision from time to time. This method is applied to the sale of timber, fire-wood, bamboo and minor forest produce. Besides this the sale from depot is also done by this method.

(3) By auction

This method is the most common method adopted for the sale of forest produce. In this method the intending purchaser fixes the price by outbidding each other. The highest bidder becomes the purchaser. To prevent unreasonably low price, a minimum price is fixed which is known as the Upset price or the Reserve price. Below this Upset price the bids are not accepted and the lots are withdrawn, if the minimum price is not obtained. This is the most profitable method of sale, if keen and healthy competition among buyers exists. This method of sale too has certain disadvantages as follows :

a) It is unsuitable for slack times when markets are overstocked.

b) There is a likelihood of the purchasers forming rings and defeating the purpose of auction.

c) This leads to unhealthy competition.

d) This method also leads to monopolisation of trade in the hands of big contractors, who outbid the small contractors.

This method though had certain defects; yet it was widely prevalent not only in Uttar Pradesh but whole of India. The contractor system of working based on sale by auction is dealt ahead in detail.

(4) Sale by tender

In this system, the intending purchasers make offers through tenders on or before fixed date, stating the price they are willing to pay. These tenders are of two types "

1) Sealed tenders.

2) Open tenders.

(1) *Sealed Tenders* - In sealed tenders, no tenderer is able to know the bid offered by the other., The date, time and place of opening these tenders is notified in advance. In it each tender is accompanied by the treasury challan of Earnest Money deposit (10% of the contract value deposited in the treasury and pledged to the competent authority).

(2) *Open tenders* - In open tenders, the purchasers send their offer before a fixed date. In this the offer of other purchasers are known. It is in fact a tender-cum auction arrangement.

Sale by tender is suitable for produce having fairly constant value and in the case when the market is slack and overstocked. This method is also used in breaking rings.

Of these various methods of exploitation and sale the contractors system of exploitation based on sale by auction was largely practiced in Uttar Pradesh forests. A detailed description of the procedure of auction of forest produce to the contractors (former pattern of working) is as follows :

Forest produce has been divided into two broad categories :

1) Major forest produce.

2) Minor forest produce.

Major forest produce include all types of timber such as industrial, constructional, agricultural and fire-wood. The timber resources are plentiful, varied and of good quality. The timber and fire wood come from coniferous and broad leaves species. Broad leaves species are mainly of hard woods, such as - Sal, Saim, Khair, Shisham, Haldu, etc. They become exploitable at various ages ranging from 20 years to 120 years.[19]

In minor forest produce, a number of products other than wood are classed, such as - bamboo, ringals, grasses, leaves, gum of semal, sal seeds, resin, honey, horns, hides and skins, lime stone, stones, boulders and bazri, a number of medicinal herbs, bark of kilmora, casuarina, acasias, etc.

The exploitation of forest produce before the emergence of U.P. Forest Corporation was done as follows :

Minor forest prodcue was sold on lump sum basis by auction except in the case of Khair lots and Bhabar grasses for which a long term leaves (contract) was given. Honey was extracted both : departmentally and through purchaser agency.

The major forest produce was generally sold on lump sum basis by public auctions without guarantee of out turn. The annual auctions were held by the forest department and good prices were obtained.

The procedure of marketing of forest produce by the forest department is as follows :

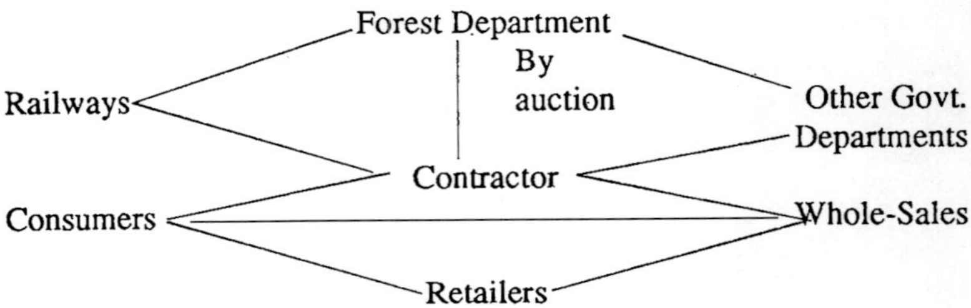

There were two methods two methods of marketing forest produce in U.P. Viz. :

i) Departmentally, and

ii) Purchasers agency through auction.

Departmentally, the forest department extracted only a few forest produce. Resin was extracted from pine from the forest of Kumaon, Tehri and Western Circles and was supplied to the Indian Turpentine and Resin Co. Ltd., Cooperative Society, Ranikhet Grah Udyog Sahkari Samiti Limited, Kotdwara, etc. Certain quantities of fire-wood within a reasonable radius from all primary stations in Bahraich, North Kheri and South Kheri Divisions continued to supply fire-wood to the towns of Kanpur, Lucknow, certain military stations, sugar factories and other industries, through approved agents at controlled rates. Besides this, the forest department also supplied its produce to the Star Paper Mill and Railways.

The forest department markets its major proportion of forest produce through purchaser's agency (i.e. contractors) by auction. The contractors on their part after purchasing the forest produce, extracts them from forest and sell them to the wholesalers (sometimes directly to the retailers also). The whole-salers sells the produce to the retailers from where it reaches the consumers. The whole-salers may also resort to selling of produce directly to the consumers. The contractors also supply to the Railways and other Government Departments directly at certain fixed rates declared at auction.

The most common method used by U.P. Forest Department to sell its produce (major forest produce) to the contractors is by means of open auction (the former pattern of working). In this, the contractor carries out the complete operation from exploitation to marketing, i.e., the contractor solely carries out the work of felling, cutting, extracting, collecting, transporting, storing and finally marketing the timber. A brief discussion of the complete process of marketing timber through contractor's system is as follows :

The forest under the control of U.P. Forest Department are worked under a systematic procedure of working based on silvicultural rules. This systematic procedure is incorporated in the working plan of the concerned forest division, which is prepared for a period of ten years. The working plans clearly indicate the species and their number and sizes which are to be marked for felling. The marketing in different forest regions are done strictly in accordance to the prescription laid down in the working plan. The Divisional Forest Officer (D.F.O.) who is the head of the administration of that division conducts the sale of forest coupes to the contractors.

The following procedure is followed in conducting the sale of timber to the contractors :

(1) Firstly, the marking of trees is done. The trees are marked strictly in accordance to the prescriptions of the working plan of the forest division. The following types of trees are marked :

a) The trees which are mentioned in the working plan prescription or which are required for some special schemes.

b) All dead, dying and diseased trees are marked.

c) Old and overmature trees are also marked.

d) Trees which are dangerous to the surroundings are marked.

e) Sometimes trees are marked under special schemes, such as constructions of dams, etc.

The trees are marked at two places -

i) At breast height level that is 4' 6" above the ground level.

ii) At 4' 5" above the ground.

If the tree is standing on the sloping ground then the measu rement should be taken from the uphill side.

The size of 'gadave' (stamp mark) is 9" x 9".

After marking of trees is over, the marking record is properly compiled, classifying species wise in 4" (10 cm.) diameter classes and grouping them under sound, fit and unfit.

(2) Second step is preparation of sale-list. The marked trees are properly scrutinized and grouped into 10 cm diameter classes and an extract of such enumeration is prepared, forming lots of suitable sizes. List of lots are then submitted by all Rangers to the Divisional Forest Officer. The Divisional Forest Officer then prepares a consolidated sale-list and sends it to the press for printing, after inviting tenders if these are not to be printed in the Government press. Simultaneously, wide publicity is given through local and other News Papers about the intended dates and places of auction, requesting the contractors to obtain printed sale-lists from the office of the D.F.O.

3) Thirdly, the Divisional Forest Officer prepares confidentially a statement of estimated price, known as Upset Price, [20] which he

expects in the auction sale. The statement of Upset Price is sent to the Conservator of Forests for the purpose of his approval and information. The Conservator of Forests examines these prices and makes alternations where necessary and then sends the list back to the Divisional Forest Officer. This Upset Price is a highly confidential matter between the Conservator of Forests and the Divisional Forest Officer, till the auction is over. The Upset Price serves as a guide-line in conducting sale, as this helps to decide whether or not the highest bid is to be accepted and it also enables the D.F.O. to raise security, if the bid goes much in excess of the confidential estimate.

(4) Fourthly, only those contractors who are enrolled with the forest department will be permitted to take part in the auction, (except in the auctions where full payment is to be made at the fall of hammer). Fee payable for enrolment and renewal is as follows:

Period--One year

	Enrolment fee	Renewal fee
Whole of U.P.	Rs. 250/-	Rs. 100/-
One Forest Circle	Rs. 100/-	Rs. 50/-
One Forest Division	Rs. 50/-	Rs. 25/-

SOURCE: General conditions of sale, U.P. Forest Department.

The period of enrolment will be from 1st July to 30th June. The renewal of enrolment should be done within one year of enrolment, otherwise a fresh enrolment will be required.

(5) Fifthly, on the day of auction, the contractors have to deposit Rs.500/- as Earnest Money Deposit before entering auction hall and are also required to furnish names of two persons (Partner or Agent) to who will accompany him to the auction. The Earnest Money Deposit is a mark of their genuine interest in participating in the auction sale. If the contractor leaves the hall without signing the sale-list or causes any disturbance in the hall then this money is liable to be confiscated.

(6) Sixthly, the auction is conducted by the Divisional Forest Officer at the specified place on the definite date. The auction is done on forest coupe (lot) wise basis (as specified in the sale-list). The contractors do the bidding for the losts which they intend to purchase. The highest successful bidder become the purchaser of the lot, at the fall of hammer of the auctioning authority i.e. the Divisional Forest Officer.

(7) Seventhly, after the bidding for the lot or forest coupe is over the following procedure is adopted :

i) The successful contractor has to sign the "Agreement Deed'' in the presence of witness. On behalf of the State Govt.the Divisional Forest Officer signs the deed.

ii) The successful bidder has to sign the bid list and deposit the complete security and Stamp Duty (according to Indian Stamp Duty Act, 1899). If he fails to do so, the sale of that lot is cancelled and it is auctioned again.

iii) If the contractor furnishes is found to furnish a wrong address then the work is stopped and the lot is sold again. If any loss is incurred then this would be recovered from the contractor giving wrong address.

(8) Eighthly, after completing the sale of all the forest coupes a consolidated report comparing the sale price and the up price along the comment of the Divisional Forest Officer is prepared. The consolidated report is then submitted to the Conservator of Forests for the purpose of approval.

(9) Ninthly, the acceptance of auction takes place as follows :

i) The Divisional Forest Officer has the power to ratify the lots which have crossed the Upset Price and are within his financial limits.

ii) The sale of lots above Rs. 5000/- are confirmed only after the approval of the Conservator of Forests.

iii) The sale of lots can be cancelled within one week of auction, if the bid value is low.

iv) No complaints relating to auction will be entertained after 10 days of auction.

The Conservator of Forests after giving approval (quashing where necessary) sends the consolidated sale-list back to the Divisional Forest Officer.

(10) Tenthly, the Divisional Forest Officer after receiving the approval, puts the sale-list on the Notice Board of the Head Office and the Range-Office. The putting up of such notice will be treated as sufficient information to the contractors. If the contractor does not receive any information of his lot within 40 days of auction, then he should find out from the forest officers. If the information of acceptance is received after 40 days then, there will be no binding on the contractor to accept such lots.

(11) Eleventhly, the sale of lots will be registered only in one name or firm (if legal registration of firm is shown and if the bidder is a legal representative of the firm). In normal course no transfer of contract from the name of one person to another person will be effected, except in the exceptional cases as follows :

i) On the death of contractor.

ii) When the contractor is incapacitated

iii) When the contractor has become bankrupt or when he is finally migrating.

(12) The following standing instructions are observed for security deposits :

(i) The minimum security deposit for a lot will be 25% of its sale value but in no case less than Rs.200/-.

(ii) If the financial position of the Contractor is not known to the D.F.O., then in such cases the security can be doubled.

(iii) If the bid of a lot becomes very high then in such a case the Divisional Forest Officer can ask the bidder to deposit 60% of the bid value, even if it is in the first bid.

(iv) When the security taken is more than 30% of the bid value, then the excess over 30% of security deposit is adjusted in the first second instalment.

(v) The security deposit can be in any form depending on thediscretion of the D.F.O. Besides cash, it can be in the form of cash certificates, defence bonds, savings certificates or fixed deposits in the State Bank of India or any other nationalised bank, pledged to the Divisional Forest Officer.

(13) The method payment of instalment is as follows :

i) The instalment of the lot will be deposited through Treasury Challan.

ii) The fixation of instalments will be done in the following manner:

a) The contractors have to deposit the full amount immediately after auction/sale, where the value of lot is upto Rs. 10,000/-.

b) For lots valuing above Rs. 10,000/- the payment will be done in two instalments as follows :

--First instalment should be deposited by the first November or before starting work (i.e. extracting/ felling), which ever is earlier. The first instalment should be 50% of the sale value of the lot or Rs. 10000/- which ever is more.

--The second instalment to be paid by 1st January is of the remaining amount or 50% of the sale value of the lot.

--If the contractor fails to pay the instalment on the due date then he will be liable to pay the late fee which will be as follows :

a--) If paid within 30 days (the grace period), no late fee will be charged.

b-) If payment is done after 30 days and within 60 days then the late fee of two paise per hundred per day i.e. 7.30% per annum, would be charged but in it the grace period is also included.

c-) If the payment or any part of payment is done after expiry of 60 days then a late fee of 5 paise/100/day i.e.18.25% per annum will be charged from the due date of deposit of instalment.

(14) After receiving the approval of forest coupe,the contractors check their lots. If he finds any shortage in his lot, he complaints about it to the D.F.O. For this the contractor has to deposit Rs. 100/-. In such cases, the recounting of trees is done and if there is any shortage then it is noted down and the value of the lot is also reduced in the same proportion (proportion of the missing trees to the total number of trees in the lot).

(15) Next step involved is the felling of trees, which is done keeping the silvicultural norms as follows :

Felling is usually done during winter season i.e. October to March (except in the hilly Himalayan Regions above 2000 meters, where it is done from April onwards, but in no case during the winter season). Felling during the rainy season is done only at certain places. The summer season is generally avoided for felling as there is every danger of fire and the logs dried too rapidly which causes splitting and cracking of wood. Winter season is best suited as the trees are quite at rest and the labour is also easily available.

The contractors employ local labour for the purpose of felling. These labour are hired labour or are supplied by the labour contractors. The contractors also employ technical hands like Munshis, Moharrirs, etc. These labourers employed in the felling area is supervised by the kith and kin of the contractors. Sometimes, the expert fellers i.e. Axe-men and Sawyers, who are imported from Bareilly, Delhi and Mandi in Himachal Pradesh are also employed. Local Sawyers like 'Shirolis' of Kumaon and 'Nagpuri' of Garhwal are also used for this purpose.

Felling of trees is usually done by the help of saws and axe in two ways :

i) By the use of axe alone.

ii) By the use of axe and saw.

Felling is done as low as possible near to the ground (but over the lower stump mark). Felling by contractors in U.P. is mostly done by the help of axe alone. The felling by axe causes considerable wastage of timber. It is estimated to be 2% to 5% or more. This happens because the fellers dislike to lean over sufficiently and secondly, high stumps are usually left in regions of snow, steep slopes and forests covered with bushes, fallen timber and debris. In recent past the use of saws was also introduced by the contractor for the purpose of felling.

After felling the tree to the ground the conversion of tree into logs is done. The process of conversion involves debarking of tree, trimming off the branches and sawing the clean bole and sometimes the larger branches into lengths of required sizes. The conversion of tree into logs by the contractors depended on the market requirements and facilities of transportation. Covnersion was carried out mostly by cross-cut saw. The logs are generally of long length of even sizes. The preferable length of log is 5 meter or multiples thereof. Length for the poles is from 3 meter to 15 meters. Where timber is required in special sizes, the logs should be cut in the forest to conform to the specifications. The conversion of timber is also done keeping in view the method of transportation to be adopted.

(16) All the converted timber should carefully be recorded before it leaves the area of exploitation. The Contractor's 'Munshis' makes a systematic record and assigns a serial number to each log. This serial number is deeply carved on the surface of the lot by means of chisel. At the same time, the Munshi's also put their property mark and year of conversion on the lot.

(17) After completing the recording, the transportation is done. For the purpose of transportation, the forest department issues the permit to the contractor known as the 'Nikasi Permit'. This permit entitles the contractor to transport the timber to any part of the country, without any hinderance coming on the way

The transportation takes place in two phases :

1) Firstly, the timber is transported from the area of exploitation to the road side or launching depot (when transportation is done by means of water).

2) Secondly, when the final transportation takes place from the road side or from the launching depot to the contractor's depots.

The various methods generally adopted by contractors for transporting timber from the area of exploitation to the road side are as follows:

a) *Carriage by men* - Generally in hilly regions, where the roads are not present, small and medium size timber like sleepers small poles and fuel wood are carried to the road side on men's shoulders or on backs.

b) *Dragging by animals* - In many parts of the forest, dragging by ponies, mules, camels and elephants is done.

c) *Sliding* - Sliding is also done in cetain hilly regions.

d) *Rolling* - Rolling of logs was adopted in areas of gentle slopes, but due to the damage which results to the logs and the forest has led to its abandonment.

e) *Tramways* - Forest tramways are usually constructed over level or fairly level ground. These are light railways of small gauge and their engines are run by fuel wood. Due to their being uneconomical, they have since been discontinued. Tram ways were used to bring timber out of the forests in Gorakhpur and Haldwani Divisions.

f) *Carriage by carts* - Carts were also used for transporting timber from felling coupes to the road side. Sometimes, when the felling coupes are near the contractor's depot, then the carts are even used to transport timber from felling coupes to the depots.

In the second phase, the final transportation of timber takes place from the road side to the Contractor's depots by the help of motor transport through trucks or sometimes even with the help of Railways wagons.

During the transportation the timber is scrutinized on site on the way and at the check-posts situated on roads.

(18) Lastly, the timber i.e. the logs are converted into sawn timber of various sizes in the contractor's depots with the help of saw-mill and then the timber is finally sold to the customers through retail trade.

The extraction of forest produce by the contractor system had been prevalent for more than 100 years, yet it had failed to develop on the modern lines. The contractors system of working had large number of draw-backs as follows:

a) The contractors never followed the scientific method of working based on silvicultural norms, thus causing great loss to the forest.

b) The contractors involved themselves in overbidding and then to compensate the loss on account of overbid, resorted to illegal felling of trees.

c) The Forest Department also faced problem in collecting the revenue of the lots. The contractors generally used to delay the payment of instalments and sometimes even refuse to do the payment. In order to collect such payments the forest department had to go for litigation, thus, making such collections uneconomic in nature.

d) Contractors continued to use traditional methods of exploitation i.e. by axe alone or by manual saw and axe. These methods used to cause loss not only to the produce but also to the surrounding forest.

e) Contractor used to exploit the forest labour. They gave the labour low wages and that too not in time. In certain cases it was observed that the contractors had even started using certain labour as bonded labourers.

f) The contractor system had failed in generating permanent employment in the forestry sector. They never make use of educated unemployed youths in their field operations.

g) The contractors neither followed the logging techniques nor made use of logging experts in performing logging operations.

h) The contractors had failed in meeting the allotted supply of the public sector and Government departments. Whenever they supplied the allotted timber, it was observed that it was of poor quality.

i) The contractors had even failed in supplying good quality timber and fire-wood to the local public at reasonable rates.

Because of these discrepancies, the State Government decided to nationalise the forest trade and introduced the new concept of State Trading in forest produce. The U.P. Forest Department abandoned the contractors system of extraction of forest produce and handed over the complete processing and marketing of forest produce to the U.P. Forest Corporation.

References

1. R.D. Gupta & D.P. Joshi : "Working Plan for the Pithoragarh Forest Division,1968-69 to 1977-78, p. 81 U.P. Forest Deptt.
2. L.C. Sharma, "The Forest Industry in U. P.", p. 391. Thesis for Ph. D., Agra University, 1972.
3. Ibid, p. 392.
4. E.P. Stebing, Forest Utilization in India, Vol. III, FRI.
5. L.C. Sharma, "Forest Economics-Planning & Management," p. 41, 1980.
6. R.C. Soni-The Centenary of Forest Administration in U.P., 1861-1961, U.P. Forest Department, 1961.
7. K. Nath : Working Plan for the Dehradun Forest Division, 1959-60-1968-69, Part I, Chapter V, p. 44.
8. L.C. Sharma : Forest Industry in U.P., p. 397, Thesis for Ph. D., Agra University, 1972.
9. A Brief History of Forest Administration in U.P. p. The Centenary of Forest Administration in U.P., 1861-1961, U.P. Forest Department, 1961.
10. A Brief History of Forest Administration, p. 19, The Centenary of Forest Administration, 1861-1961, U.P., Forest Department, 1961.

11. L.C. Sharma-Forest Industry in U.P., p. 398, Thesis for Ph. D., Agra University, 1972.
12. L.C. Sharma - " " p. 403, " " "
13. A Brief History of Forest Administration in U.P. p. 19; The Centenary of Forest Administration, in U.P. 1961.
14. H.C. Dey : Forest Administration in Kumaon, p. 53. The Centenary of Forest Administration in U.P., 1861-1961, U.P. Forest Department, 1961.
15. A Brief History of Forest Administration in U.P. p. 19. The Centenary of Forest Administration in U.P., 1961.
16. G.N. Singh : Forest Administration & Public Opinion, p. 44, The Centenary of Forest Administration in U.P., 1861-1961, U.P. Forest Department. 1961.
17. A Brief History of Forest Administration, p. 19, The Centenary of Forest Administration in U.P., 1861-1961, U.P. Forest Department, 1961.
18. J.S. Chauhan, "Forest Resources of Kumaon Region of U.P. and their Exploitation", p. 189-Thesis for Ph. D. Agra Univ., 1975.
19. L.C. Sharma, 'The Forest Industry in U.P.', p. 214. Thesis for Ph. D., Agra University, 1972.
20. Upset Price =4/5 (Market Price)-Cost of Production.

CHAPTER THREE

Emergence of U.P. Forest Corporation - Backdrop and Factors

Forestry sector has a much greater potential both in regard to production and employment generation. But this generation of production and employment requires a planned forestry development on appropriate lines. The planned forestry development will increase the volume of production per hectare and will also fetch more value from forest products. Such a forestry development will support the economy of the backward areas and the tribal population which depends largely on the forests. A good deal of thought was given by the foresters to the intensive management of the forests and forestry development.[1] Food and Agricultural Organization of the United Nations in its paper (Unasylva, Vol. 23 (4) No. 95. 1969) on production forestry stressed that, '

'production forestry is a business, and that organization of forest service must, therefore, differ in some important respect from that of other government departments in which the duties are limited to provision of services.'[2]

Food and Agricultural Organization emphasized that in order to enable the forestry sector to develop to its full potential, it is very necessary to do the following: [3]

1) Forestry and forest industries administration should be given a relatively extra governmental and more business like character by making it a corporation.

2) More means should be found for integrating the growing side of forestry with the industrial side. To this end, it may be advisable for governments either to control forest industries or to participate in their ownership and management.

3) Specialists on the field of forestry, development planning and non-wood services of forestry should be assigned to the ministries or other units that are responsible for the national or regional development planning. They should, however, remain members of the forest service, and be ultimately responsible to the head of that service.

4) The administration of forestry departments should be guided not by the general body of rules and regulations which govern the civil services but by a special set of rules and regulations which take into account the characteristics of the forestry and forest industries sector, and the business nature of government forestry.

5) The distribution of field personnels should be rationalised special attention being paid to the input/output relationships that are involved in the development of staff in different areas.

6) Above all, "the nation must be development oriented, and governments must provide forest administration with the necessary funds to perform their manifold duties. This may mean a special dispensation with regards to the time-honoured system of allocations through annual budgets."

All these various points emphasized by the Food and Agricultural Organization aimed towards the adoption of a commercial approach to forestry development. These suggestions were considered by various other countries of the world. The experience of these countries showed a positive effect/trend. They were of the opinion that 'forestry can give higher returns per hectare provided intensive management was adopted including creation of man-made forests on a far more larger scale that had been possible for lack of adequate investment.

The National Commission on Agriculture was also convinced with the views of Food and Agriculture Organization. That is why the Commission agreed to adopt a new commercial approach to forestry development to get higher returns per hectare. But all this required intensive management and the creation of man-made forests on a far more larger scale which requires higher expenditure per hectare in inputs and labour than was usually required for conservation forestry. The National Commission on Agriculture, therefore, recommended a change over from traditional approach of conservation forestry to a more modern dynamic programme of production forestry.[4] This modern dynamic a programme of production forestry required a large scale investment in the forestry sector, provided that such investment should contribute its due share to the generation of national income and the creation of employment. It was very difficult for the state budget to provide such large funds. Hence, the National Commission on Agriculture recommended recourse to the institutional financing. But, the flow of such institutional-finance was only possible if there is a company or a corporation to take the funds, as these financial institutions can not lend to the state governments. The Commission also ruled out the joint sector approach in the industrial sphere of forestry. The National Commission on Agriculture identified Agricultural Refinance and Development Corporation as the prime institution for financing long-term forestry programmes.[5]

Based on the recommendation of Commission, in the year 1968, an experimental study was done by Maharashtra Forest Department by forming a Forest Development Board. Certain selected areas were entrusted to the Board by the Maharashtra Forest Department to accelerate the pace of forestry. The primary objective of the Board was to increase the productivity of the miscellaneous slow growing low yielding forests by clearfelling and planting them commercially valuable species like teak and bamboos. The Board was given complete autonomy in its working. The Board completed the work successfully and showed a positive trend in increasing the yield.[6] The National Commission on Agriculture, seeing the excellent work done by the Maharashtra Development Board came out with the recommendation to establish the Forest Development Corporations as a public sector in the country to handle the work of production forestry in India.

Public sector enterprise in India has recently emerged as a dominant partner in the nation's economy and has been assigned a wide role to play in the process of economic development. The pattern and personality of the public sector is inevitably conditioned and influenced by the ethos, political philosophies, historical background, traditional usages and customs and by geographical and natural conditions. In almost all the important central sectors of production and industry, large corporations have been set up. The States also picked up the example of the centre and set up several corporations and support institutions and organizations. The establishment of corporation for undertaking any activity in an efficient manner seems to have become the fashion of the day.[7]

The main thesis of National Commission on Agriculture was that the contribution of forestry sector to the national product has been extremely negligible. The contribution from forestry to the Gross Domestic Product and to the employment generation has been 1.6% and 0.2% respectively, which is quite insignificant. The national annual average gross revenue per hectare from India's productive forests, under the prevailing system of working was only about Rs. 21.50 per hectare and employment was only 0.2% of rural working force.[8] The forestry sector has a much greater potential both in regard to production as well as employment. National Commission on Agriculture laid stress on intensive management of forests including creation of man-made forests on a far larger scale than was hither-to possible for lack of adequate investment. The Commission recommended that these practices require higher investments per hectare in inputs and labour than what was normally required by conservation forestry. The Commission also gave an example of agricultural sector which got out of the rut of low production only when scientific management with balanced inputs at high cost was introduced. All the above mentioned findings lent a strong recommendation to the formation of public sector in the field of forestry.

The public sector concept in the forestry sector is a late entrant and came into sharp focus mainly as a result of the recommendation of the National Commission on Agriculture in the year 1972.[9] This recommendation was followed by a strong recommendation made by the First Forestry and Forest Products Conference held at Dehradun in 1973 (Recommendation No. 3), wherein it was recommended to set up Forest Development Corporation for raising, harvesting and sale of industrial wood.

The concept of Forest Development Corporation was thus evolved by the National Commission of Agriculture with the following specific objectives[10] of :

1) making use of institutional finance,

2) ploughing back substantial portion of profits for development,

3) giving flexibility and freedom to managers to introduce innovations and take quick and timely decisions, and

4) to diversify production and to induct into the organization experts from related disciplines from outside.

The Forest Development Corporations were assigned the following activities :

a) to harvest hither-to inaccessible mixed quality forests transferred to it, by adequate investments on roads and machineries.

b) to plant, grow, cultivate, produce and raise plantations of selected species to meet the requirements for pulpwood, industrial timber, constructional timber, etc., tailored to the need of the state and economy.

c) to promote forest industries to utilize the wood released as a result of intensified feeling programme and utilization of subsequent planted material.

d) to promote companies, firms, establishment, concerns or undertakings for the purpose of development of industries based on forest produce and to assist and finance any individual or company with capital and credit resources.

The Commission also suggested that the land and the standing trees, which are transferred to the Forest Development Corporation are to be valued and these two values together should form the base of equity capital of corporation against which it can borrow from lending institutions. In addition, the State Government grants additional equity for enabling the FDC to ensure adequate investment on roads, machineries, etc., in formation years. The Central Government also used to contribute an amount equivalent to the cash equity contribution of the State Government, if the Forest Development Corporation fulfilled certain conditions. But this was stopped in 1980.

The Forest Development Corporations were expected to have following advantages over the department [11]:

a) Diversification of activities which would otherwise be difficult for the department under the government.

b) Linking the production programme with marketing and utilization. Risk-taking in the field of marketing.

c) Greater capacity and scope for experimentation.

d) Undertaking tribal and rural welfare activities.

e) Development of local skills, arts and crafts.

The report of National Commission on Agriculture- "Interim Report on Production Forestry - Man made Forests, August 1972" could be considered a major land mark in Indian Forestry. Among the major recommendations of the National Commission on Agriculture was that pertaining to the establishment of Forest Development Corporations in the States (each State is expected to have a Forest Development Corporation). This recommendation was received favourably by the Government of India and the States and the result was that in all sixteen Forest Development Corporations were set up in different states. These Forest Development Corporations were established as Registered companies under the Company Act, with an exception of Uttar Pradesh Forest Development Corporation (U.P.F.D.C.) which formed as a statutory body, by an act of legislation.[12]

A study of the Articles and Memorandum of Association of Forest Development Corporations reveal that all possible and imaginable activities pertaining to forests and forest based industries have been included in them. It may take many years for the forest Development Corporations to achieve a sizeable part of the objectives set out in these documents. Basically, in accordance with the recommendations of National Commission on Agriculture, Forest Development Corporations were primarily meant to undertake the task of raising man-made forests and setting up forest based industries.[13]

The Forest Development Corporations as a group exhibits considerable diversity in their strategies, objectives and operational goals. These diversities are mainly due to ecological and demographic features prevailing in each state and due to their financial structures. Though, they have diversity yet, in general, all the Forest Development Corporations follow the 'National Commission on Agriculture' model. The classification of the Forest Development Corporation, on the basis of their repective activities, is as follows:

(1) Corporations working like agents to the State Government

Maharashtra Forest Development Corporation is an agent of the State Government. Initially, the State had a Forest Development Board or undertaking plantation activities. Later on this Board was converted into a Corporation. The Board worked in forests on an agency basis. This idea was continued in the Corporation. But, as the Maharashtra Forest Development Corporation worked as an agent to the Government, hence, this restricted the borrowing capacity of the Corporation as ARDC had refused to extend institutional finance. [14]

(2) Corporation Performing Extended Functions of the Forest Departments viz. Exploitation, Logging, etc.

Uttar Pradesh Forest Development Corporation is performing extended function of the Forest Department. It is statutory body with no equity base. The Corporation is assigned the role of a contractor for doing logging operation and marketing of felled forest produce. Recently, some areas have been transferred to it. The Corporation has set up its saws mills for sawing logs obtained from forest areas. The State Government gave the corporation a sum of nearly Rs. 2.5 million at the time of formation to commecne its operation. After that, the Forest Development Corporation was expected to have a cash credit arrangements with banks.[15]

(3) Corporation Undertaking Exploitation and Marketing Works in Remote and Inaccessible Regions

West Bengal Forest Development Corporation comes in the category. But, recently this corporation has also entered the industrial field.[16]

(4) Corporations Dealing in Exclusively Minor Forest Produce

Gujarat Forest Development Corporation was primarily set up to collect and market tendu leaves for Beedi Industry. The Corporation later on took up the collection of Achar, Gum, Mahua flowers and seeds, etc. This Corporation has also planned to, diversify its activities by setting up an integrated wood working unit and managing problematic forest areas.[17]

Similarly, Bihar and Orissa Forest Development Corporations are also doing sizeable trade in Sal seeds.

(5) Corporations Taking Plantation work on Lands on which the Forest Department has felled and removed forest produce.[18]

Karnataka Forest Development Corporation is doing plantation of Eucalyptus mainly for supply of raw material to Rayon units on the forest land transferred by Forest Department after removal of crop - I.

(6)Corporations Undertaing Plantation either for specific Industries or for Economic Production[19]

Examples of such Forest Development Corporation are Kerala, Tamil Nadu, and Andhra Pradesh. Kerala Forest Development Corporation is raising Eucalyptus and Cardamum plantation. There is a separate corporation for raising rubber plantation. Andhra Pradesh Forest Development Corporation is concerned with raising plantation of Bamboo, cashew and coffee. Orissa has also started plantation of casuarina and cashew in the coastal areas.

(7) Corporations Setting Up and Managing Forest Industries[20]

The Himachal Pradesh Forest Development Corporation is running a resin and turpentine factory. The Andaman Forest Development Corporation is running plywood/veneering units in a joint sector with some private units. Karnataka has a separate entity from KFDC by the name of Karnataka Forest Industries Corporation.

(8) Corporations with Diversified Activities like Plantation, Industrial Development, Research and Development, etc.

Madhya Pradesh Forest Development Corporation is unique in this respect. It performs following activities [21]:

a) Raising teak and bamboo plantation.

b) Establishment of wood-working units.

c) Establishing cottage and small scale industries.

d) Undertaking research and development activities.

e) Undertaking intercropping operation on a large scale.

Based on the diverse activities performed by the various Forest Development Corporations, their respective State Governments have made different arrangements for the transfer of forest area. In Maharashtra, total project area are transferred to the Forest Development Corporation. In Kerala, only annual plantation area, after removing timber and plywood material, are handed over to the Corporation. In Karnataka, the plantation area is tranferred to the Corporation after removing Crop - I. In Uttar Pradesh, entire divisions are transferred to the Corporation. And lastly, in Madhya Pradesh, areas sufficient for 10 years plantation activity are transferred to the Corporation.

As regard the question of payment of lease rental by Corporations to the State Government, it has already been pointed out that the recommendation of the National Commission on Agriculture (N.C.A.) was for payment of annual rental equivalent to the revenue foregone from the transferred areas. Each State adopted its own idea with regard to the lease rental question. In West Bengal, the F.D.C. pays the Government lease rental alongwith royalty. In Maharashtra, all the revenues are made over to the State Government. In Uttar Pradesh, the Corporation pays royalty to the State Government on the basis of average of previous 3 years market price. The prevailing figures of lease rental etc. in some F.D.C's in relation to the gross revenue are as follows [22]:

West Bengall	–	11%
Tamil Nadu	-	13%
Uttar Pradesh	-	30%
Bihar	-	40%
Madhya Pradesh	-	19% (Expected tobe come around 40%).

The National Commission on Agriculture examined the role played by the forest contractors (former method of working) in the exploitation of forests. The commission came to firm conclusion that the forest contractor system of exploitation of forest resources has been greatly harmful not only to the forests and the State Department but also to the people involved in forestry operations i.e. the forest labourers. The Commission in its final report, published in 1976, gave the following findings :

a) It recommended the elimination of the contractor system of working. altogether from forestry operation, in the interest of better revenue to the State Exchequer and the fair wages to the workers employed in forestry operation. The contractors were motivated by self interest, hence they often operated in disregard of rules and also sometimes jeopardised the basic requisite of forest conservation.

b) Litigations and malpractices in the forests is an annual headache to the State Forest Department.

c) The losses should be eliminated which is due to pilferage and skin methods to which the contractors are largely accustomed.

d) The wages given by the contractors to its forest labour are very less.

e) The private forest growers of forest produce or farmers owning forest products do not get reasonable market price.

f) The forest contractors are enjoying huge profits from forests. This is causing concentration of economic power among the forest contractors. This is done at the cost of State Revenue, which is adversely affecting the pace of forestry development.

g) The Forest Departments are facing the problem in the collection of royalty/revenue from the contractors. This forces the forest department to go for action.

h) The contractors sometimes indulge in excess bidding of forest coupe and to overcome their loss, they indulge in over-exploitation of forest resources.

i) The contractors also lower the bid value by forming association or cartels. This lowers the revenue of the State Forest Department.

j) The Forest Contractors are indulging in illicit feeling and illegal removal of forest produce.

k) The contractors are using traditional methods of exploitation of forest produce, which is harmful to the surrounding forest produce.

Based on the above findings, the National Commission on Agriculture recommended the abolition of the forest contractor system of exploitation and recommended the formation of Forest Development Corporation to replace the forest contractor system. This will help in increasing the direct interest in the utilization of forest and will also help in increasing the yield from forest i.e. the commercial value of forests. According to the National Commission on Agriculture Report, 1976,

> "No timber should be sold standing in the forest. Only after departmental logging, either through direct employment or through labour cooperatives, all timbers should be brought to sale depots outside the forests and the sale should be made at that point only."[23]

The Uttar Pradesh Government studied the findings of the National Commission on Agriculture in the context of U.P. Forest Department.

It was observed that the above mentioned findings were quite true to the State Forest Department. Based on the findings of the National Commission on Agriculature, the Uttar Pradesh Government accepted the recommendation of the Commission and formed the Uttar Pradesh Forest Development Corporation. The Uttar Pradesh Forest Development Corporation was constituted through an Ordinance on 25 November 1974.[24] The U.P. Forest Development Corporation was assigned the role of a contractor for doing logging operation and for marketing the felled forest produce. The Ordinance was later replaced by the U.P. Forest Corporation Act (U.P. Act, No. 4, 1975)[25]. The U.P. Forest Corporation is constituted as a 'Local Authority' very much unlike other Forest Development Corporations, which are established as Registered Companies under the Company Act.

The Uttar Pradesh legislature passed the U.P. Forest Corporation Act, 1974 (U.P. Act No.4, 1975) for the establishment of the Corporation (U.P. Forest Corporation) for a better preservation, supervision and development of forests and better exploitation of forest produce within the State and matter connected with forests. According to the U.P. Forest Corporation Act, 1974, the U.P. Forest Corporation was constituted on 25th November 1974, by the State Government Notification in the Gazette. The Corporation was constituted as a Corporate Body having perpetual succession and a common seal. It shall have the power to acquire, hold and dispose of property and to sue and to be sued in its corporate name for the purpose of the Act.[26]

The Head Office of U.P. Forest Corporation is located in Lucknow, with the branch offices at other places in the State where considered necessary. The Corporation was constituted as a 'Local Authority' with the following constitution -

(A) The Corporation will consist of a Chairman and 8 other Members. All of them were to be appointed by the State Government, U.P. and their appointments shall be notified in the Gazette. Out of these eitht Members, five were to be appointed from amongst the Officers serving under the State Government. One of these five members shall be appointed as the Managing Director of the Corporation. Beside these five members, not more than three of non-official members were to be appointed by the U.P. Government from amongst the persons who in its opinion possess experience relating to preservation and development of forests. Before making the appointment it was to be made sure that they should not have any disqualification which could be detrimental to the interest of the Corporation or the General public.27

A person shall be disqualified to be appointed as Chairman or member, if he :

i) has been convicted of an offence

ii) is an undischarged insolvent

iii) is physically or mentally handicapped

iv) is having share or interest in sale, purchase, lease or exchage of immovable property, etc.

(B) The term of office of the Chairman and non-official member will be generally three years unless it is determined earlier by the State Government by Notification in the Gazette. They can also resign from office by submitting his resignation in his hand writing to the State Government.

(C) The Corporation can appoint such employees which it considers necessary for the efficient performance of the corporation. But such appointment should be done in consultation with the Public Service Commission, U.P.

(D) The employees of the Corporation shall be entitled to draw such salaries and allowances from the fund of the Corporation as determined by regulations.

Subject to the provision of the U.P. Forest Corporation Act, Chapter III, and to any general or specific directions of the State Government, the Corporation will perform the following functions -

a) to undertake removal and disposal of trees and exploitation of forest resources entrusted to it by the State Government.

b) to prepare projects relating to forestry and to undertake research programme relating to forest and forest products.

c) to render technical advice to the State Governments on matters relating to forestry.

d) to manage, maintain and develop such forests as are transferred or entrusted to it by the State Governments.

e) to perform such functions as the State Government may from time to time requires.

The aims, objectives and functions of the U.P. Forest Corporation are discussed in detail in the following chapter No.4.

The U.P. Forest Corporation was vested with the following powers

i) to set up workshops or factories for processing forest raw material;

ii) to establish, maintain and operate laboratories and experimental and research stations;

iii) to borrow money, issue debentures and manage its funds;

iv) to enter into such contract or arrangement with any person as the Corporation may deem necessary for performing its function under this Act;

v) to do anything which may be necessary or expedient for carrying out its functions;

vi) to incur expenditure and grant loans and advances to such persons as the Corporation may deem necessary for performing its functions.

According to the provisions laid down in the Chapter IV of the Act, the Corporation will have its own fund, which shall be a local fund and to which shall be credited all money received by or on behalf of the Corporation. This fund will be used to meet the expenses incurred by the Corporation in discharging of its functions. The money of the fund shall be kept in the State Bank of India or in the Uttar Pradesh Cooperative Bank or in any Scheduled Bank.

The U.P. Forest Corporation may with the previous sanction of State Government can borrow any sum required for the purpose of this Act, whether by the issue of bonds or stocks or by making arrangements with the bankers. The State Government may also come to help the Corporation by providing subsidy or aid. The State Government may also advance loans to the Corporation on certain terms and conditions as decided by the State Government. For the purpose of repayment of any loan raised by the Corporation, the Corporation will establish a sinking fund. The Sinking Fund shall be maintained, invested and applied in such manner as may be perescribed.

The Corporation will make the budget of its estimated receipts and expenditure of the next financial year ensuing. The Corporation was to maintain proper accounts and get its audit done annually by the Examiner, Local Fund Accounts or any persons authorised by him. The certified report by the Examiner, Local Fund Accounts was to be forwarded annually to the State Government.

The Corporation at the end of each financial year shall prepare and submit a report to the State Government giving an account of its activities during the previous year and of the activities which are likely to be undertaken by the Corporation in the next financial year. The State Government will lay such report before each House of the State Legislature.

All the operations of the U.P. Forest Corporation are executed through the Managing Director under the direction of the Governing body constituted under the U.P. Forest Corporation Act. The Managing Director is the Chief Executive, appointed directly by the Government. The Governing Body comprises of a Chairman and eight members (including the Managing Director) out of which three are non-official and the rest are Government Officers. The Managing Director is assisted by General Managers, Accounts Officers-cum-Financial Advisers, Audit Officers, Marketing Managers, Logging Managers, Sales Manager and similar other officers. Similarly, a large number of class three and four staff for harvesting, marketing, accounting, etc. have been recruited from time to time under the norms laid down by the Governing Body. In the initial stage, all the officiers were on deputation from the U.P. Forest Department. But for the last five or six years, it has become difficult for the Forest Department to spare adequate number of suitable

Forest Officers for the Corporation. In such a situation the Corporation started building up its own cadre of class two technical officers from the year 1981-82. The Corporation is also recruiting Logging Officers directly. In course of time these logging officers after promotion will man the post Managers. For the time being these posts will be filled by on deputation from Forest Department.

The Corporation is also making arrangment to provide training in forestry and logging techniques. The training is being provided at For st REsearch Institute, Dehradun for Officers. For class III Assistant Logging Supervisors/Logging Supervisors training arrangements are done at Rangers College, Haldwani through U.P. Forest Department. Besides this, Officers are required to attend short term training courses at Logging Development Institute of Government of India at Dehradun.

A detailed study of Organizational structure of the U.P. Forest Corporation is done in Chapter - V.

The U.P. Forest Corporation has completely taken over the responsibility of exploitation and marketing of forest produce in U.P. In the field of exploitation, it is making use of modern techniques of production to avoid wastage of forest produce. It is also ensuring working based on modern logging concepts through a trained fleet of experts. It is also making use of better transportation technique. A complete study of exploitation, U.P. Forest Corporation.

The Corporation after processing brings the produce to the sale depots from where the marketing is done. Presently, the Corproation is marketing its produce through following methods:

1) By Public Auction.

2) Through Private Treaty.

3) Through Retail Sale.

4) Direct Supplies to the Government and Public Sectors.

In order to do better marketing of its products the Corporation has opened 102 sales depots. Yet, it is not able to perform the marketing properly. A detailed study of Marketing of Forest Products by U.P. Forest Corporation is done in Chapter VII

Finance

The U.P. Forest Corporation has been constituted as a 'Local Authority' and that is why it is precluded. from having share capital. It can borrow from the State Government and it can issue debentures and stocks. The Corporation was launched with a modest loan of Rs. 15 lacs from the Government. Initially, the loan was for one year but later it was extended to 3 years. The loan of Rs. 15 lacs was paid back by the Corporation to the State Government alongwith interest in 1976-77. In the y car 1975-76, the Corporation also received a cash-credit arrangement of Rs. 20 lacs from the State Bank of India.

During the coming year, the Corporation constantly expanded its activities by ploughing back the profits earned and by arranging cash credit facilities from the bank. The State Bank of India increased the cash credit limit to Rs. 2.35 lacs and the Bank of Baroda also provided cash-credit of Rs. 60 lacs. Besides this, the Corporation also rceived loans of Rs. 15 lacs from Hill Development Corporation and Rs. 50 lacs from Kumaon Mandal Vikas Nigam.

It was only in the initial years, the U.P. Forest Corporation made use of the loan and cash-credit facility. The Corporation made has repaid back all the loans and cash credit alongwith interest. It is since 1982-83, the Corporation is not using the cash credit facility provided by the banks.

Audit and Accounts

The accounting year of the Corporation has been kept from 1st October to 30th September. This has been done as the field work in forest generally starts in November, after the rains end. The audit of accounts of the Corporation is done by the Examiner, Local Fund and by a firm of Chartered Accountants.

Progress of U.P. Forest Coporation [28]

In January 1975, the Corporation obtained a loan of Rs. 15 lacs from the State Government to meet its working capital requirement for taking up logging operation in three ranges of South Gonda, 3 ranges of South Pilibhit and one lot of Uttar Kashi Forest Division. Thus, only three logging division (including the departmental logging unit at Uttar Kashi) were opened in the inaugural year. The total timber production was 0.26 lacs cu.meters and the toal turn over was above Rs. 64 lacs. Thereafter, about 3 logging divisions were created every year till 1981-82, as more and more forest areas were allotted to the Corporation for logging work. The Manager-in-charge of logging division also looked after the disposal of forest produce upto 1976-77. But this arrangement was not found satisfactory and thereafter, a separate Sales Organization was created in the Corporation in 1977-78. The sales Organization was headed by Dy. General Manager, assisted by three Sales Officers. Thus, in the year 1981-82 in all 20 logging divisions became operative to harvest all lots of Tehri Garhwal and Kumaon Circles, Siwalik circle (excluding Eucalyptus lots of Bijnor forest division), all lots of Kalagarh Forest Division of Western Circle, all lots of Central and Eastern Circle, excluding Soth and North Gorakhpur Forest Divisions. The total timber production in that year was 4.79 lac cu.m. and the total turn over was Rs. 34.26 crores.

In the year 1982-83, the Corporation took over harvesting of entire major forest produce and some important items of minor forest produce of all Government forest excluding Social Forestry areas. Thus, all timber, bamboo, cane, baib grass, honey-wax and tendu leaves lots were harvested by the Corporation. About 6.31 lac cu.m. of stacked firewood, 7 lacs scoere of bamboo and 0.11 lac metric tonnes of baib grass produced. About 8,000 cu.m. of soft wood was supplied to various industrial units. This year, it was for the first time that the Corporatuin also harvested tendu leaves with the cooperation of Forest Department. About 3.90 lacs standard bags of Tendu leaves was produced. The labour was paid 6 paise per bandle of 50 (fifty) tendu leaves fit for Biri manufacture. Besides this, as an incentive 35,000 Dhotis were distributed free of cost among the labour who were also supplied wheat flour at subsidised rates. The total turnover in the year 1982-83 was Rs.59.25 crores. Similarly, in the year 1983-84, 5.88 lac cu.m. of timber, 7.94 lacs cu.m. stacked of fire-wood, 5.15 lacs score of bamboo, 3.98 lacs standard bags of tendu leaves and 0.7 lac metric tonnes of Baib grass w as produced and total turnover was Rs. 76.24 croes. In the year 1984-85, 5.89 lacs cu.m. of timber, 4.88 lacs cu.m. stacked of firewood, 4.65 lac score of bamboo, 3.40 lacs standard bags of tendu leaves and 0.10 lac metric tonnes of Baib grass was produced. The total turnover is Rs. 91 crores. In 1985-86, 6.58 lacs cu.m. timber, 5.40 lacs cu.,m stacked fire-wood, 5.84 lacs scores of bamboo, 3.84 lacs scores of bamboo 3.80 lacs standard bags of tendu leave and 0.89 lac metric tonnes of Baib grass was produced. The total turnover of the Corporation increased to Rs. 105.09 crores. The above progress in the field operations and in the turnover reflects that the Forest Corporation has developed at very fast pace and has been able to achieve the objectives for which it was formed to a great extent.

The creation of U.P. Forest Corporation has brought about a marked change in the ethos and concept of forest performed by the Forest Department have given place to those of resource management in a business like manner with the full involvement of labour, which is now treated as a part of total system and not as an extraneous element. A high priority has been assigned to the creation of man-made forests to meet hte escal;ating demand of the people and industries for timber, poles and fuelwood. This would be borne out by a comparison of average outlay in relation to gross revenue on plantation in the Forest Department with The outlays of Forest Corporation. The percentage outlay in Forest Department was anything from 0.5 to 0.8 per cent, while in Corporation it is from 10 to 15 per cent. With regard to the question of employment generation, the intensive operations of U.P. Forest Corporation lead to a twelve fold increase in employment in comparison with conventional management practices obtaining in the Forest Department. This creation of employment in forestry will certainly help in curbing unemployment.

The U.P. Forest Corporation has no reached a stage where it can diversify its activities and expand its working in coming years. The Hon'ble Chief Minister has also emphasized on various occasions that the Corporation should not limit its activities to harvesting and market-ing should not limit its activities to harvesting and marketing of forest produce only, but that it should also take up development activities such as plantation of forest areas and intallation of industrial units based on various major and minor forest produce and other ancillary units to cater to the various requirements of the forest department e.g. manaufacture of polythene bags, barbed wire, resin tapping tools and other equipments etc.

The Corporation has also formulated a 10 year corporate plan (1986 - 1995) in order to identify new fields of activities and diversification to be undertaken by the Corporation. Broadly speaking the corporate plan will comprise the follwoing fields of activities [29]:

1) to harvest and to market forest produce in commercially exploitable Government forests, in social forestry areas in the plains and Civil Soyam, Panchayat, Cantonment and Municipal forest in the hills;

2) to arrange marketing of timber from tree planted on private lands in villages under the extensive trees plantation;

3) to plant up species of industrial importance, such as Eucalyptus through a subsidiary unit of Corporation or in a joint ventrue with a private consumer industrial unit;

4) to establish small scale industries (such as manu facture of hard board and briquettes from pine needles, carpentary work based on secondary timbers, processing and marketing of honey, manufacture of polythene bags, etc.).

It is quite obvious from the above facts that the Corporation is fully alive to its responsibilities and it is continuously making study progress inspite of various constraints and problems. Itg is facing criticism from all quarters, yet, the U.P. Forest Corporation has now a definite place under the sun. The Corporation is playing a pivotal role in a development of forestry and is also accelerating its growth.

Refrences

1. A.R. Maslekar, "Managing the Forests", pp. 128-129, 1983.
2. Unasylva, Vol. 23 (4) No. 95, 1969.
3. Unasylva, Vol. 23 (4) No. 95, 1969.
4. V.K. Seth, "Forestry Corporations in India : Present Status and Future Strategies", P. 5, Second Forestry Conference, F.R.I., Dehradun, October, 1979.
5. Report of the National Commission on Agriculture; Part IX, Forestry, 1976. Government of India.
6. A.R. Maslekar, "Managing the Forests", pp. 130-131, Dehradun, 1983.
7. Report of National Commission on Agriculture; Part IX, Forestry 1976, Government of India.
8. Report of the National Commission on Agriculture, Part IX, Forestry, 1976. Government of India.

9. National Commission on Agriculture, "Interim Report on Production Forestry-Man Made Forests", August 1972, Govt. of India.
10. A.R. Maslekar, "Managing the Forests", p. 131, 1983.
11. A.R. Maslekar, "Managing the Forests", 1983, p. 132.
12. V.K. Sethi. "Forestry Corporations in India-Present Status and Future Strategies", P. 4, Second Forestry Conference, F.R.I. Dehradun, 1979.
13. *Ibid*, p. 11.
14. V.K. Seth, "Forestry Corporations-Present Status & Future Strategies", p. 11. Second Forestry Conference, F.R.I., Dehradun, 1979.
15. V.K. Seth. "Forestry Corporations, Present Status and Future Strategies", p. 12. Second Forestry Conference, F.R.I., Dehradun, 1979.
16. *Ibid*, p. 12.
17. *Ibid*, p. 12.
18. V.K. Seth, "Forestry Corporation-Present Status & Future Strategies, p. 12, II Forestry Conf., F.R.I., Dehradun, 1979.
19. *Ibid*, pp. 12-13.
20. *Ibid*, p. 13.
21. V.K. Seth, "Forestry Corporations-Present Status and Future Strategies", pp. 13-14, Second Forestry Conference, F.R.I., Dehradun, 1979.
22. V.K., Seth, "Forestry Corporation-Present Status & Future Strategies", pp. 14-15, Second Forestry Conference, F.R.I., Dehradun, 1979.
23. Report of the National Commission on Agriculture, Part IX, Forestry, 1976.
24. U.P. Forest Corporation-A Short Note on its Activities, p. 1, 1974-75 to 1984-85.
25. The U.P. Forest Corporation Act, 1974, p. 1, Lucknow 1975.
26. U.P. Forest Corporation Act, 1974, Chapter II, p. 2.
27. *Ibid*, pp. 2-3.
28. Based on Annual Report of U.P. Forest Corportation.
29. Annual Report; U.P. Forest Corporation, 1985-86, pp. 19-20.

CHAPTER FOUR

AIMS, Objectives and Functions of U.P. Forest Corporation

Food and Agriculture Organization of the United Nations emphasized in the paper, "Production Forestry" (Unasylva, Vol. 23 (4) No.95, 1969) that in order to enable the forestry sector to develop to its full potential, it is very necessary to give the administration of Forestry and Forest Industries to relatively extra government and more business like character, like Corporation. In pursuance to the above findings, the National Commission of Agriculture submitted its "Interim Report on Production Forestry--Man-made Forests" in August 1972. The National Commission on Agriculture in its report recommended the establishment of Forest Corporation.

It was only on the recommendation of National Commission on Agriculture that the Uttar Pradesh Forest Corporation was constituted through an Ordinance on 25th November 1974. This Ordinance has since been replaced by the U.P. Forest Corporation Act (U.P. Act, No.4, 1975).[1] The U.P. Forest Corporation was constituted as a "Local Authority", very much unlike other Forest Corporations, which are established under company act.

The U.P. Forest Corporation Act (U.P. Act No.4, 1975) laid down certain provisions and directions under which the U.P. Forest Corporation was to perform its duties. Subject to the provisions of this Act and to any general or special directions of the State Government, the

U.P. Forest Corporation has the following specified functions to perform[2], namely-

(a) To undertake the removal and disposal of trees and exploitation of forest resources entrusted to it by the State Government,

(b) To prepare projects relating to forestry within the State,

(c) To undertake research programmes relating to forest and forest products and to render technical advice to the State Government on ma tters relating to forestry,

(d) To manage, maintain and develop such forests as are transferred or entrusted to it by the State Government,

(e) To perform such functions which the State Government may from time to time require.

Out of these aforesaid functions the U.P. Forest Corporation has so far been entrusted by the State Government only to perform the function of harvesting and marketing of forest produce. In the beginning, the U.P. Forest Corporation was alloted logging work only in the problematic forest areas which were notorious for illicit fellings and other serious problems. Gradually, year after year, more and more forest areas were allotted to the Corporation for logging work. By the year 1982-83 and thereafter, the entire State Government Forests (excluding Social Forestry Areas) were allotted to the U.P. Forest Corporation for logging work.

In the context of the above laid down function under the Act, the U.P. Forest Corporation started functioning to achieve the following objectives [3]:

a) To eliminate labour contractors from field operations such as felling, billeting, logging (conversion), carriage (off road transport), operation, etc. in the forests and to do all these works itself.

b) To use developed/improved logging implements for felling and logging purposes, so that the wastage may be minimise and the production of cut and sawn timber may be maximised.

c) To constitute forest labour societies and to engage them in works of the forest to the maximum.

d) To ensure payment of minimum wages to the forest labourers, get for them cheap and best rations, and to end up their exploitations at the hands of forest contractors.

e) To make available and supply good quality timber to the Government Departments and Public Sector Industries.

f) To prevent damage caused by fellings and carriage in the forests and to use modernised techniques by eliminating the use of old obsolete traditional techniques.

g) To set up wood based industries and to meet the demand of wood of the existing industries.

h) To make available good quality wood to the common people at reasonable rates.

i) Gradually to end the contractor system of working from the whole State and to get the entire work of the forests done by the Corporation.

j) To set up industries based on forest.

k) To reforest (plant) the felled area.

l) To construct roads for working in the remote forest areas on hills.

m) To set up such small industries in backward forest areas which may procure means of livelihood to the local people and also to make available school and medical facilities.

n) To engage educated unemployed youths in forestry operations after giving them training in modern logging techniques.

The U.P. Forest Corporation is performing its functions efficiently. The Corporation has achieved its functions to a very great extent. A brief account of the achievements of the objectives of the U.P. Forest Corporation is given as follows :

a) The U.P. Forest Corporation has been able to eliminate the labour contractors from the work of logging operations. It has now completely undertaken the responsibility of felling, logging, transportation, storing and finally marketing the forest produce entrusted to it by the State Government. By undertaking the marketing and harvesting operations, the U.P. Forest Corporation has helped in solving the serious problem of illicit felling.

b) The U.P. Forest Corporation has discarded the use of old techniques of production and has brought modern methods of production in its working. The Corporation has abandoned the use of axe, traditional manual saws, etc., and has substituted these equipments by mechanical devices like advanced manual saws, mechanical saws, etc. This shift in technique has helped in checking colossal wastage at the time of fellings, which were inherent in the old system.

c) The U.P.. Forest Corporation has stopped the use of traditional system of transportation which was towing or dragging. This system used to damage the younger forest crop. The U.P. Forest Corporation has abandoned the use of this system and has started using gravity ropeway and skyline crane. This has helped in checking the destruction which was caused during conversion and transportation of wood.

d) The U.P. Forest Corporation has largely helped in checking the exploitation of labour at the hands of forest contractors. The exploitation of labour use to take place in the form of payment of less wages and more work, with practically no facilities. The U.P. Forest Corporation has made it a point to remove the evils of the contractor system of working. The Corporation ensured the payment of minimum wages to the labour as laid down by the Government. In fact the wages are actually paid to the forest labour on piece rate basis in certain remote forest areas and are much higher than the prescribed minimum wages.

e) The U.P. Forest Corporation helped in the formation and development of forest labourers cooperative societies. These labour cooperatives were given all types of assistance. Even in certain hilly areas, forest labour is supplied ration at subsidised rates and this subsidy amounts to Rs.3/- to Rs.4/- per labour per day. Besides this, the Corporation also provided the housing facilities.

f) The U.P. Forest Corporation has encouraged the use of scientific methods of extraction. This helped in bringing modernisation and has also increased the efficiency of the labour. Now the labour is able to do the work in less time and with less discomfort. This increases his efficiency and thereby increases the income of the labour. The increase in income helps in raising the standard of living of the labour.

g) The U.P. Forest Corporation has helped in specialization by providing basic logging training to its labour. During the time of training the Corporation also provided allowances to the labour. Such training helped in fostering the development of forest corporation on modern lines.

h) The U.P. Forest Corporation has also performed the duty of supplying good quality timber to the Government departments and Public sector industries at reasonably cheap rates. The main institutions to which the Corporation supplies timber and fire wood, etc. are Railways, B.H.E.L., U.P. S.E.B., U.P. State Rajkiya Nirman Nigam, Sugar Mills, Border Road Directorate, Garhwal Mandal Vikas Nigam, AGRO, Hindustan Aluminium Corporation, P.W.D., Irrigation and other Government Departments. Besides this, the Corporation also supplies raw material (soft wood) to various private sector industries as per the allotment made by the Government.[5]

i) The U.P. Forest Corporation is also supplying fire-wood at cheap rates on ration card to local domestic consumers in various towns located in the districts in which harvesting of forest crops is done by the Corporation. Good quality timber is also supplied to the local public for their bonafide use at reasonable retail rates from various sale depots.

j) The U.P. Forest Corporation has also engaged the educated unemployed youths in its forestry operations. It has also made necessary arrangements of training in modern logging methods.

k) The U.P. Forest Corporation has also made use of solar kiln for drying wood. It has also taken into its control the wood drying industry which was under the control of U.P. Small Industries Corporation.

Besides the above direct benefits, the creation of Forest Corporation has also resulted in providing the following indirect benefits.

a) It has helped in the enhancement of payment of sale price it time and has helped in avoiding delays which generally use to happen with the contractors. Thus the formation of U.P. Forest Corporation has safe-guarded the realization of dues i.e. price of tree.

b) In the past, unnecessary rise in bids use to take place due to personal rivalry. Thus, there was always a probability of having loss. To recover this loss, they used to indulge in illicit fellings in the forest. Due to the creation of U.P. Forest Corporation, there is no such danger.

c) The Forest Corporation has also undertaken the collection of Tendu leaves which has solved the problem of bonded labour in this area.

d) The Corporation has also started developing its own nurseries to take plantation of trees.

Though the Corporation has achieved a number of objectives in a short span of time yet it has failed to perform certain functions. It has failed in making projects relating to forestry and has not been able to start research programmes relating to forestry and forest based industries. The Corporation has also to develop small scale industries based on forest resources. Besides these limitations, still it would not be wrong to conclude that the Corporation has helped the Forest Department, Forest Labour, Consumers, Forest based industries and the societies.

References

1. The U.P. Forest Corporation Act, 1974, p. 1, Lucknow, 1975.
2. U.P. Forest Corporation Act, 1974, Chapter - III, Functions and the Powers of the Corporation, pp. 6 - 7.
3. U.P. Forest Corporation, Annual Report, 1977-78, pp.1 - 2.
4. U.P. Forest Corporation, A Short Note on its Activities, 1974-75 to 1984-85, p. 10.
5. U.P. Forest Corporation, A Short Note on its Activities, 1974-75 to 1984-85, p. 5.

CHAPTER FIVE

Organization/Structure of U.P. Forest Corporation

An organization is an efficient way of satisfying a great many human wants. An organization, compared to an individual, can assemble and co-ordinate relatively large magnitudes of specialized resources under it for achieving specific objectives. An organization carries on a number of activities to be able to deliver the final produce and to be able to survive. These numerous activities can be put as follows1:

a). Activities designed to ensure the availability of resources, personnel, material, finance, natural as well as resources of the market.

b) Organization ensures a proper flow of these activities and keeps them under control.

c) Organization performs such activities which help in main-taining the public relations and assist in building the image of the institution.

d) Organization performs activities to keep the institution in the state of dynamic equilibrium.

All these activities include the intergration of the needs of the organization, needs of its members and of other institutions related to it, with the motive of problem solving. Thus, organizations are said to be the assemblage of inter-acting human beings.

An efficient organization has following properties[2]:

i) Hierarchy of authority,

ii) Rules, procedures, controls and techniques,

iii) Formality of communication,

iv) Specialization of functions and division of labour,

v) Employment of skilled personnels,

vi) Specificity of purpose.

A sound organisation, having adequate extension methodology is needed for a successful implementation of a programme or a project of the institution. Every organisation functions in its own environment. An organisation which is most befitted in a certain set of condition or in a certain project may ultimately fail in another project or set of condition not because there was anything wrong in the project but because the people who were responsible for carrying the project did not possess adequate training and skill. Thus to make the project successful the selection of the personnel to man the organization should be carefully done.

The organisation and its environment acts and reacts and tries to achieve harmony in a given set of conditions. The functioning of an organisation is generally affected by the following factors[3] :

a) The organisation is influenced by the nature of its environment both internal and external. Internal environment is made up of a set condition which originates within the organisation, like limitation in the availability of skilled personnel, lack of finance, lack of co-ordination within the organisation, etc. External environment is made up of a set of events, information, constraints, opportunities, threats and contingencies that originate outside the organisation.

b) Second important variable affecting the forest organisation is time. A static organisation will be hazardcus in a dynamic world as it will not be able to clearly foresee the distant future thereby giving birth to risk and chances of high loss. Besides this, there is also a time-lag between the creation of the forest and full realization of its benefits.

c) Thirdly, the style and philosophy of the key decision makers (i.e. the top management) has far reaching consequences on an organisation.

d) Fourthly, Organisational structure, work-flow and technology are the means by which goals of the organisation are pursued.

e) Fifthly, organisation consists of people who react not to objectively real structures or events, but to their subjective perception.

f) Sixthly, organization is a living system with a built-in homeostatic mechanism, and it constantly adjusts its functioning in order to survive.

g) Lastly, organization is a system of inter-connected parts. A change in any one of these parts affects rest of the parts.

Thus, a sound principle of having an efficient organisation is necessary. Johnston and his co-authors[4] suggested that there should be centralization of direction but decentralization of execution. Human history shows a trend of increasing specialization. Forestry has however, been much slower in specialization than most industries. This is because the decision maker or the top-management, under whom the whole organisational structure is centralised, is concerned, with everything that affects the operational area. He is expected to have intimate knowledge of silviculture, the long-range view points of the planners, the skill of the administrator and the talent of successful business man.

Johnston and his co-authors give three reasons for slow rate of specialization in the field of forestry. First, organised forest management has evolved to a large extent in State Forest Services and Government Industries which intended to operate under somewhat less stringent economic pressures than others. That is why, urgency to adopt the most efficient method was not felt. This restricted specialization and division of labour to a large extent.

Secondly, both the aspects viz-of failures and achievements in forestry are often slow to manifestation and the advantages of changing the established stechniques are difficult to evaluate and visualise, because of five characteristic features inherent in forestry namely:

(1) A long-gestation period is required to produce its product,

(2) The timber product is also he timber growing machine which ultimately becomes product,

(3) The timber grows very slowly, thereby showing a modest rate of return,

(4) The variable cost includes a major part of growing cost of timber,

(5) Timber is versatile.

All these characteristics show that there is a time-lag, which is a hinderance to specialization.[5]

Thirdly, a great deal of forestry has been undertaken in under-developed counteries having poor communication and limited resources; conditions which neither justify nor facilitate specialization.

These reasons given by Johnston and his co-authors reveal that forestry management is of unsepcialized manner. In an unsepecialized forestry service managerial responsibilities are allocated territorially with junior managers as incharge of smaller areas of forest. Each manager is incharge and responsible for all the operations within his charge, and that is why he is expected to be jack-of-all trade. It is also true that as an enterprise becomes more specialized, an increasing proportion of the staff are given functional (specialized)responsibilities[6].

Johnston and his co-authors are of the firm opinion that there should be only three level of management-Head-quarters, district and forest. Head-quarters should be primarily responsible for direction, district for planning and finally the third level i.e. forest for implementation of the direction of the headquarters. In this method of management, each level is considered as a reducing gear which slows down the speed and increases the friction of management.

Forest organization in India is over a century old. Almost all forests are state owned. National level policies and planning is done by the central government, while implementation and administration of the forests rests with the State Government.

At the centre formerly forestry was under the ministry of Agriculture. Forestry was handled by a wing, headed by Inspector General of Forests, who is also an ex-officio-Additional Secretary to the Government of India. He is assisted by a number of Deputy and Assistant Inspector-General of Forests, A Secretary Forestry Commission, A Project Economist and Technical Assistant. A post of Additional Inspector General of Forest was recently added. It is very recently, that forestry has been transferred to the newly formed Ministry of Forest, Wildlife and Environment.

The forest department at the State level is normally under a separate department, headed by the Forest Secretary. The executive head of the forest department is the Principal Chief Conservator of Forests. He is assisted by a number of territorial as well as functional officers at different levels.

To manage the forest an efficient forest management is required. The scope of forest management is classed into three main groups as follows:

1) Control of growing stock.

2) Extraction, processing and marketing.

3) Administration of forest property.

Of these three groups, control of growing stock and Administration of forest property is done by the forest department while the specialised activities like, extraction, processing and marketing is now dealt separately by the U.P. Forest Corporation.

The Uttara Pradesh Forest Corporation was formed in the year 1974, as a statutory body. In the beginning the Chief Conservator of Forests, U.P. was also the ex-officio chairman of the corporation. The executive head of U.P. Forest Corporation was the Managing Director. All the operations of Uttar Pradesh Forest Corporation were conducted/ executed through the Managing Director under the direction of the Governing Body. The Managing Director is the Chief Executive and is appointed by the Government. The Governing Body consisted of eight members (including the Managing Director). Out of these eight members, three are non-officials and the rest five are all officers serving the Government. Managing Director is appointed from among these five official members of the Governing Body.

Managing Director of the U.P. Forest Corporation is assisted by General Manager, Chief Accounts Officer-cum-Financial Adviser, Personnel Manager, Marketing Manager, Senior Accounts Officer, Regional managers, Administrative officers, Planning and Evaluation Officer, Internal Audit Officer, Accounts Officer, Divisional Logging Manager, Divisional Sales Managers, Logging Officers and Sales Officers. Besides these, there are large number of class three and class four field staff for the purpose of harvesting and marketing[7].

There has been a massive extension in the activities of the Uttar Pradesh Forest Corporation from time to time. This extension of activities of the Corporation led to an increase the responsibilities of the Corporation. To discharge these entrusted responsibilities, the U.P. Forest Corporation has always endeavoured to reconstitute and expand. The expansion of the organization of the U.P. Forest Corporation from time to time is studied herewith.

In the beginning, the strength of the organization was less, as the corporation was a new body and it worked only in three division. The total strength of staff on 30th Sept. 1975 is as below :

a) Officers on deputation	-	11
b) Field Staff :	-	
i) On deputation	-	35
ii) Directly recrulted	-	NIL

c) Office Staff:

i) On deputation	-	6
ii) Directly recruited	-	6

d) Class IV Employees:

i) On deputation	-	9
ii) Directly recruited	-	4
Total	=	71

Source: U.P. Forest Corporation, Annual Report, 1974-75, p. 5.

From the above table on the strength of organizaation, it is evident that initially the Corporation was a very small organization and that to constituted not by its own direct employment but by the staff of forest department on deputation.

By the year 1977 - 78, the U.P. Forest Corporation had further expanded its activities to other parts of the State. This caused the creation of an independent section for sales and marketing. The appointment of Deputy General Manager (Sales) was done to look after the sales and marketing section. He was assisted by three Sales Officers and one Chief Marketing Officer. Besides this, three new posts of the Chief Logging Officers were created at Haldwani, Dehradun and Tons, who were to be assisted by Logging Officers, Deputy Logging Officers and Assistant Logging Officers. The post of Cost Accountant was upgraded to Chief Accounts Officer, who was to be assisted by an Accounts Officer. The following chart No.2 shows the organizational structure in 1977 - 78.

In the year 1978-79, the Corporation further increased its logging and sales operations in the hilly divisions of Tehri-Garhwal, Kumaon and Nath-Gonda divisions. To cope with this increasing work some reshuffling and expansion of the organization was done as follows:

CHART - I

ORGANIZATIONAL STRUCTURE OF U.P. FOREST CORPORATION, 1974 - 75 :

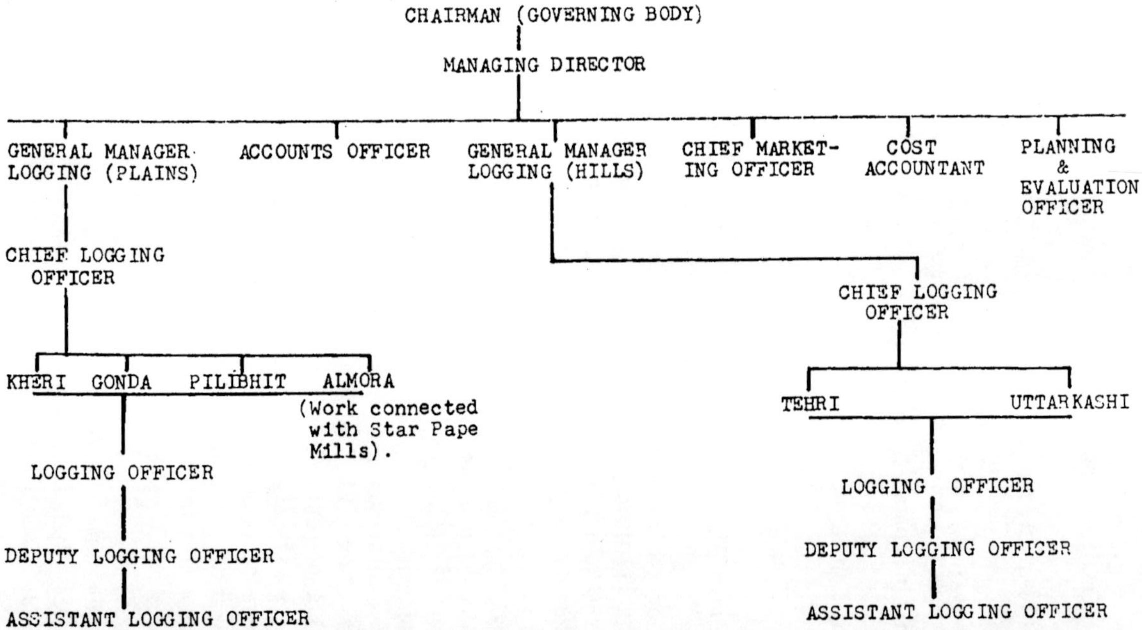

SOURCE : U.P. Forest Corporation, Annual Report, 1974-75, p. 53.

C H A R T - II

ORGANIZATIONAL STRUCTURE OF U.P. FOREST CORPORATION 1977-78 :

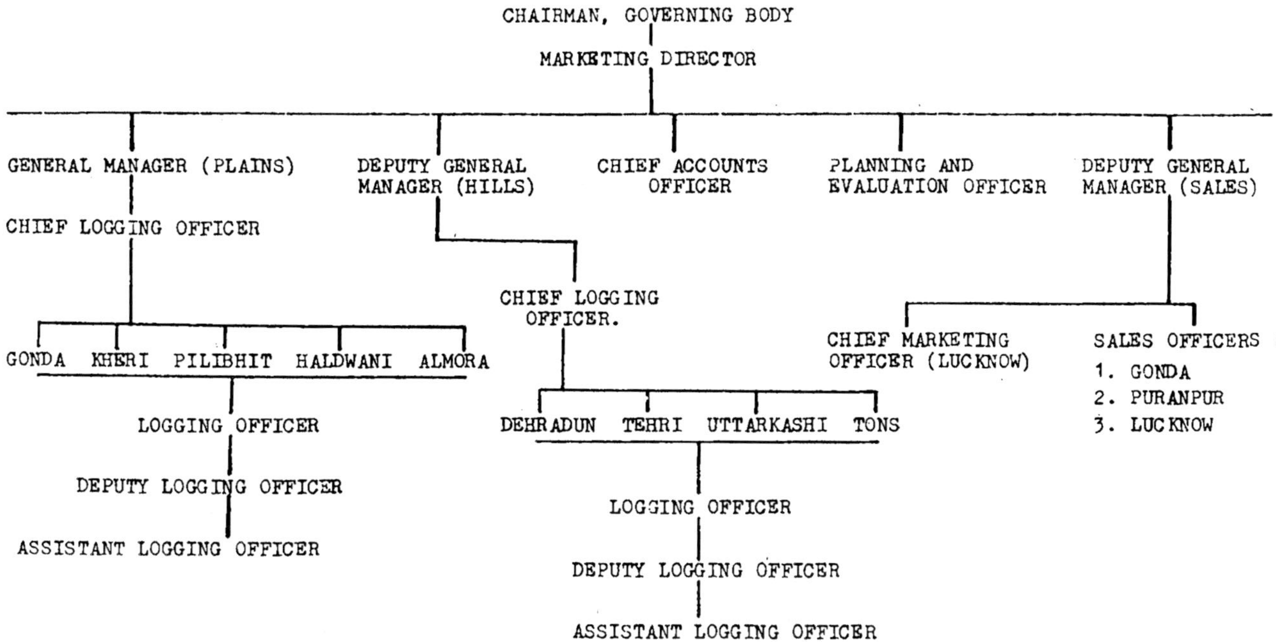

SOURCE : U.P. Forest Corporation, Annual.Report, 1977-78, p. 20.

a) The post of Chief Logging Officer was named as Divisional Logging Manager, and 8 more Divisional Logging Managers were appointed at Badrinath, Chakrata, Yamuna, Garhwal, Nainital, Ranikhet, Bageshwar and Pithoragarh.

b) The post of Chief Marketing Officer was renamed as Divisional Sales Manager and the new post of Divisional Sales Manager, Dehradun was created. The post of Sales officer, Lucknow was abolished and in its place the Sales Officer, Haldwani came into being.

c) The special post of the General Manager, U.P.F.C., was created.

d) The post of General Manager (Plairs) was abolished and instead of this, three Deputy General Managers were appointed as Deputy General Manager, Kumaon; Deputy General Manager, Plains; and Deputy General Manger, Haldwani.

e) One post of Internal Audit Officer was created.

f) One post of Administrative Officer was formed who was to be assisted by a section officer.

The overall organizational structure during 1978-79 is represented in the following organizational chart III.

The U.P. Forest Corporation went further ahead in specialisation of its activities. In order to achieve this, it further enhanced its organization by creation of the following posts :

a) A post of the Marketing Officer was formed under the Deputy General Manager (Sales).

b) Five posts of the Accounts Officers were created each under a Deputy General Manager.

c) Three Divisional logging Managers were appointed at Kalagarh, Baharaich and Lansdowne.

d) Three posts of Sales Officers were further added at Kheri, Raiwala and Kotdwara.

C H A R T - III

ORGANIZATIONAL STRUCTURE OF U.P. FOREST CORPORATION, 1978-79.

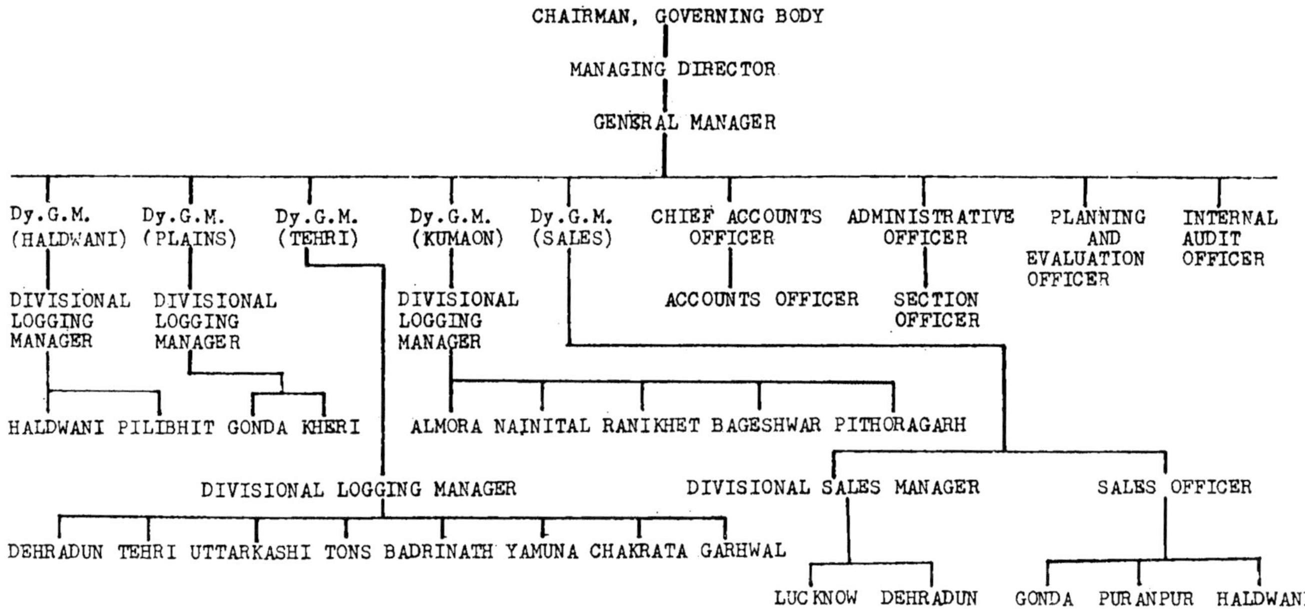

SOURCE : U.P. Forest Corporation, Annual Report, 1978-79, p. 7.

The extended organizational structure of the U.P. Forest Corporation for 1979-80 is clearly depicted in the following Chart - IV :

During 1980-81, the U.P. Forest Corporation again expanded and amended its organization in the following manner:

a) The post of Divisional Logging Manager, Bahraich was broken up into two posts: The Divisional Logging Manager, East Bahraich and the Divisional Logging Manager, West Bahraich.

b) A new post of the Deputy General Manager, Garhwal was formed probably replacing Deputy General Manager, Haldwani.

c) The Post of Deputy General Manager (Sales) was also bifurcated into two posts of Deputy General Manager, to be designated as Deputy General Manager (Sales), Eastern and Western regions.

They were to be assisted by the Divisional Sales Managers, Accounts Officer and Sales Officer as before.

d) The post of Sales Officer Kheri was abolished and instead two posts of Sales Officers-one at Bahraich and the other at Khatima-were created.

e) A new post of Statistical Officer was introduced meeting the long felt need for a separate officer for collection, complication and analysis of data, etc.

f) For the first time a post of Sleeper Passing Officer was created for inspection of the timber to be supplied to Railways and the D.G.S. & D., etc.

A graphical picture of the organizational set up during 1980-81 is as shown in the following chart - V :

In the year 1981-82, a complete reshuffle in the organizational pattern was done as given below:

a) The post of General Manager was divided into two: viz. The General Manager (Plains) at Lucknow and the General Manager (Hills) at Dehradun.

CHART - IV

ORGANIZATIONAL STRUCTURE OF U.P. FOREST CORPORATION 1979-80 :

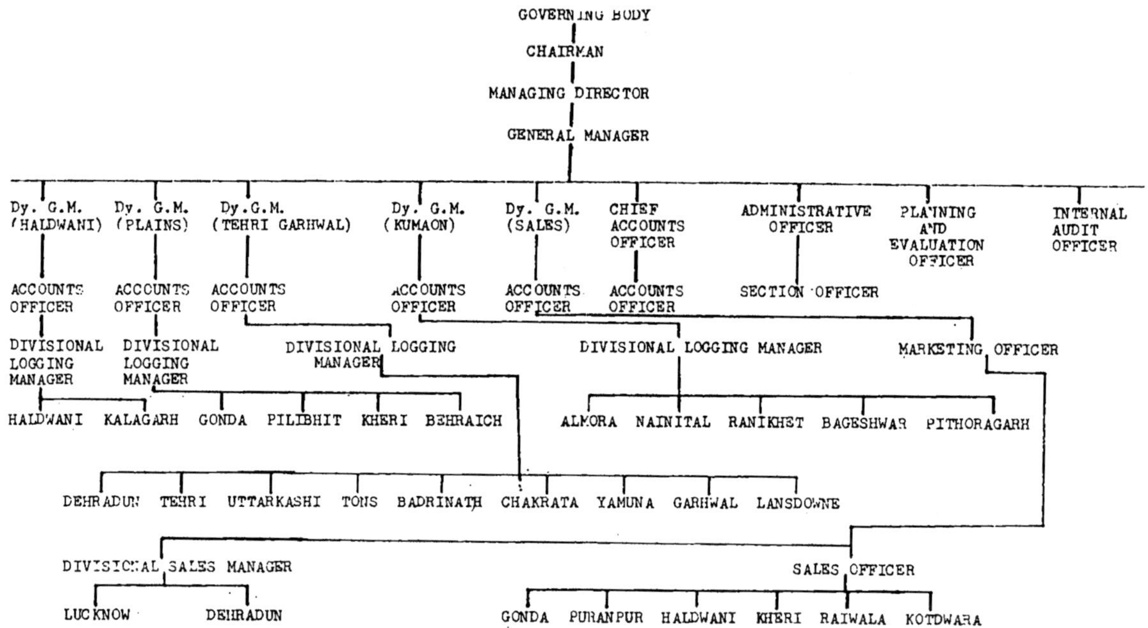

SOURCE : U.P. Forest Corporation, Annual Report, 1979-80, p. 7.

CHART - V

ORGANIZATIONAL STRUCTURE OF U.P. FOREST CORPORATION - 1980 - 81 :

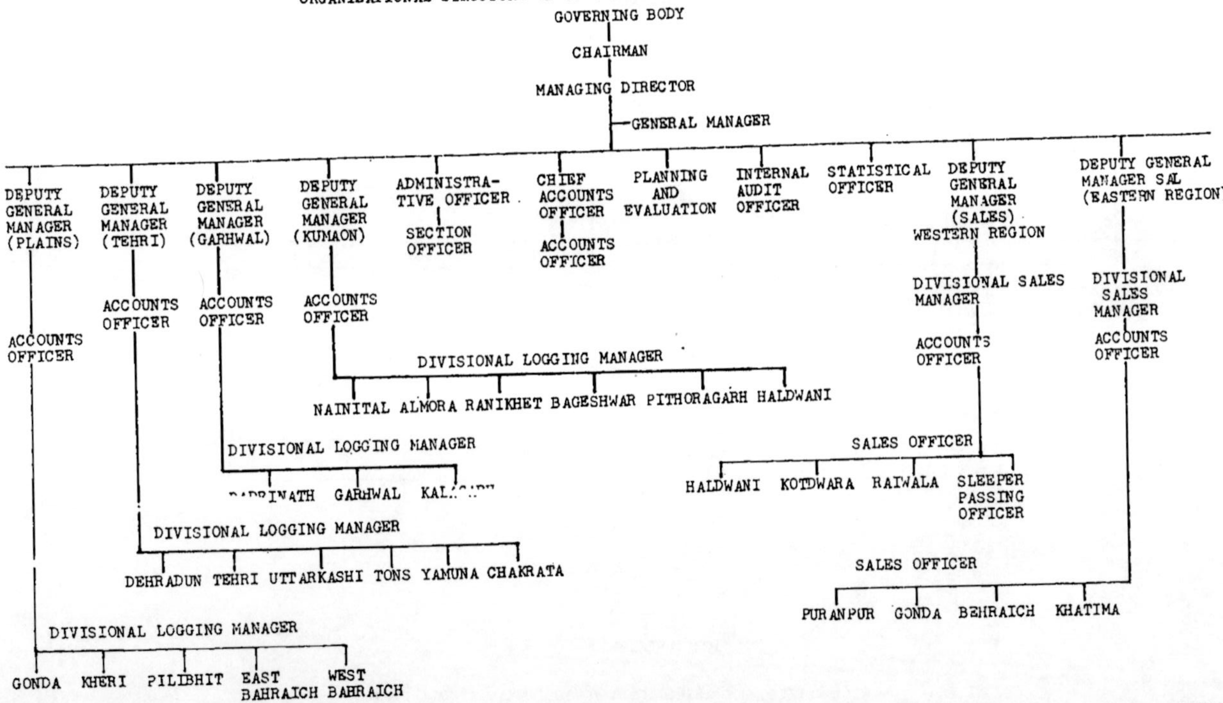

SOURCE : U.P. Forest Corporation, Annual Report, 1980-81, p. 2 (A).

b) Under each of the aforesaid General Managers, three posts of Regional Managers were created as follows:

i) General Manager (Planis) was to be assisted by Regional Manager (South) at Allahabad, Regional Manager (East) at Gorakhpur and Regional Manager (Central) at Lucknow.

ii) General Manager (Hills) was to be assisted by the Regional Manager, Kumaon at Nainital, Regional Manager, Garhwal at Kotdwara and Regional Manager, Tehri at Dehradun.

c) Regional Manager were assisted by the Divisional Sales Managers (D.S.M.) and by Divisional Logging Managers (DLM).

d) Five posts of Divisional Logging Managers (those at Garhwal, East and West Bahraich, Ranikhet and Bageshwar) were abolished and fourteen (14) new Divisional Logging Managers were appointed viz. at Robertganj, Renukut, Karwi, Lalitpur, Nath Gorakhpur, South Gorakhpur, Ramnagar, Kaladhungi, Shiwalik, Purola, Bijnor, Karnaprayag, Tanakpur, and Bahraich.

e) The posts of Divisional Sales Manager, Raiwala and Sleeper Passing Officer were abolished and 13 new posts of the Divisional Sales Managers were created viz. at Robertganj, Renukut, Karwi Lalitpur, Nichland, Gorakhpur, Lucknow, Tanakpur, Lalkua, Ramnagar or Kalagarh, Dehradun and Chutmalpur.

f) The post of the Regional Manager, Head-Quarters was also created. He was to be assisted by the Chief Accounts Officer, the Administrative Officer, the Planning and Evaluation Officer, the Accounts Officer and the Internal Audit Officer.

The completely reshuffled and newly formed organizational structure of the U.P. Food Corporation is best exhibited in the following chart - VI.

Further specialization was injected into the organization of the U.P. Forest Corporation during the Year 1983-84 to induct better managerial skill. The salient features of modified pattern of the organization are as follows:

C H A R T - VI

ORGANIZATIONAL STRUCTURE OF FOREST CORPORATION, 1981 - 1982 :

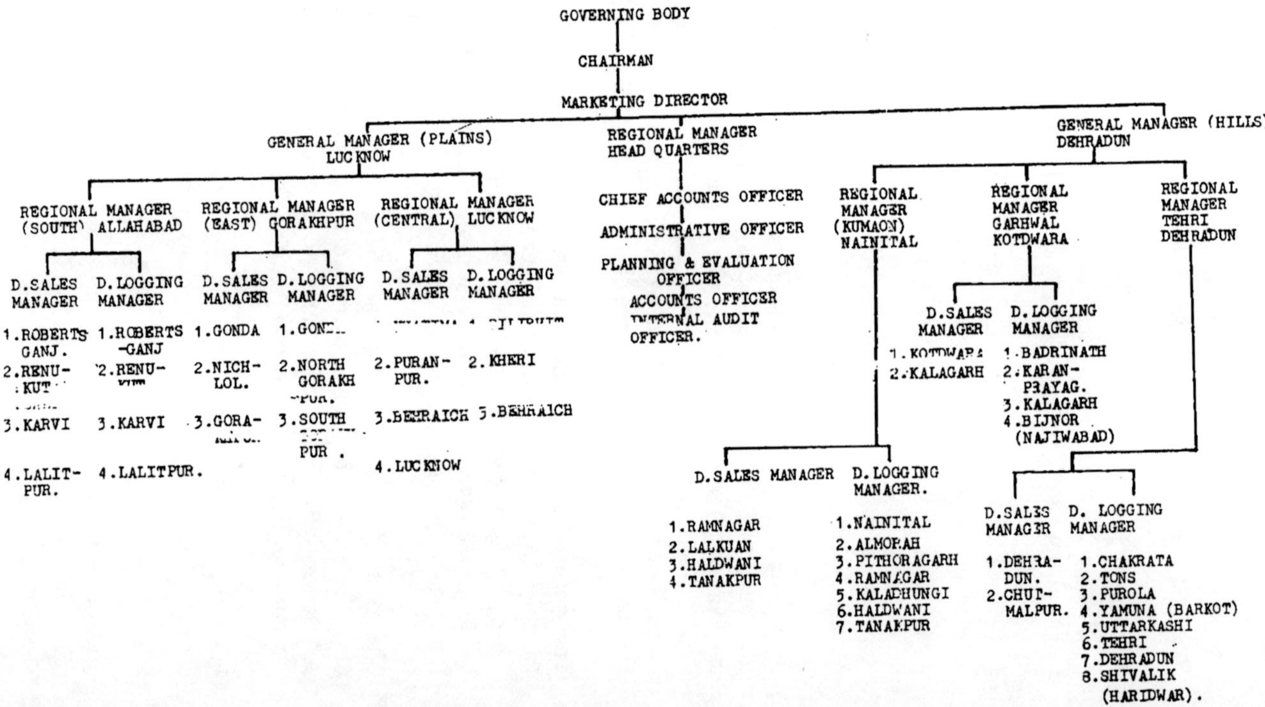

SOURCE : U.P. Forest Corporation, Annual Report, 1981-82, p. 2 (A).

a) The working of the U.P. Forest Corporation was divided into two separate branches, namely the field work and the Headquarters. This was a radical change.

b) At the headquarters, a new post of General Manager (Production) was created, who was to be assisted by the Regional Manager Headquarters and the Planning and Evaluation Officer.

c) The post of Chief Accounts Officer was upgraded as that of Financial Advisor-cum-Chief Accounts Officer.

d) A new post of Marketing Officer was created under the Marketing Manager.

e) A new post of Divisional Logging Manager, Obra was formed and each of the posts of Divisional Logging Managers at Gonda and Haldwani were broken into two viz. D.L.M.S. North and South Gonda and D.L.M.S. East and West Haldwani.

f) A new post of Divisional Sales Manager, Lalkua was created and each of the two posts of Divisional Officers, Gonda and Haldwani were bifurcated into the posts of D.S. Ms North and South Gonda and D.S. Ms North and South Gorakhpur.

g) In place of the Divisional Logging Manager, Robertsganj, a post of Divisional Logging Manager at Mirzapur was created.

The modified organizational pattern of U.P. Forest Corporation is depicted in the following chart - VIII.

A slight modifications were affected in 1984-85 in the organizational structure of the U.P. Forest Corporation, which are mentioned below:

a) The post of Statistical Officer which was abolished in the year 1981-82 was revised.

b) The posts of the Divisional Logging Manager, Baharaich, which was bifurcated into two in the year 1983-84 was amalgamated into one.

CHART - VII

ORGANIZATIONAL STRUCTURE OF U.P. FOREST CORPORATION, 1982 - 83.

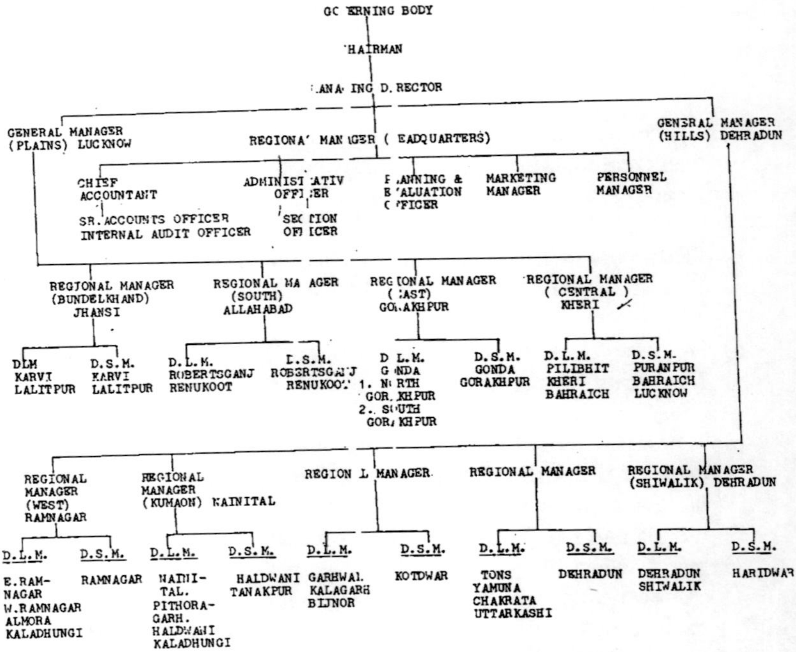

NOTE - D.L.M. = DIVISIONAL LOGGING MANAGER. D.S.M. = DIVISIONAL SALES MANAGER.

N = North S = South E = East W = West.

SOURCE : U.P. Forest Corporation, Annual Report, 1982-83, Annexture (1), p.14.

C H A R T - VIII

ORGANIZATION CHART OF U. . FOREST CORPORATION, 1983 - 84.

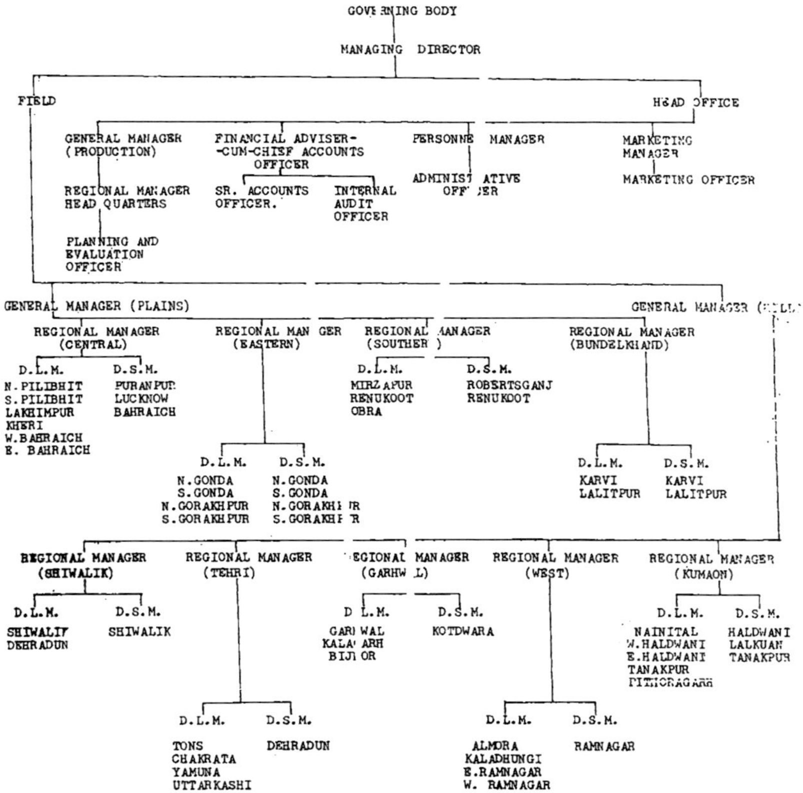

NOTE : D.L.M. = DIVISIONAL LOGGING MANAGER.

D.S.M. = DIVISIONAL SALES MANAGER.

E. = EAST. S. = SOUTH

W. = WEST N. = NORTH

SOURCE : U.P. FOREST CORPORATION ANNUAL REPORT, 1983-84. ANNEXURE (1), p. 12 (A).

c) The post of Divisional Logging Manager, Kaladhungi was bifurcated into two as D.L.M. East Kaladhungi and D.L.M. West Kaladhungi.

d) The post of the Divisional Logging Manager, Gonda, which was bifurcated into two in 1982-83 was amalgamated into one.

e) A new post of the Divisional Logging Manager, Jhansi was created.

f) Two new posts of Divisonal Sales Manager at Kaladhungi and Kheri were created.

g) Two posts of Divisional Sales Manager, Robertsganj and Renukoot were abolished and a new post of Divisional Sales Manager, Allahabad was created.

h) The two posts of Divisional Sales Managers: Karvi and Lalitpur were abolished and a new post of Divisional Manager, Jhansi was created.

The final and upto date organizational structure of the U.P. Forest Corporation at present is shown in the following chart-IX:

The year 1985-86 witnessed slight changes in the organizational structure of the corporation. These modifications were as follows:

a) The post of the Divisional Logging Manager, Pithoragarh was abolished.

b) The post of Divisional Logging Manager, Almorah which was under the Regional Manager (West) was shifted under the Regional Manager (Kumaon).

c) Nine new posts of the Accounts Officers under each Regional Manager were created.

The modified organizational structure of the U.P. Forest Corporation is depicted in the following chart -X :

CHART - IX

ORGANIZATION CHART OF U.P. FOREST CORPORATION, 1984 - 85.

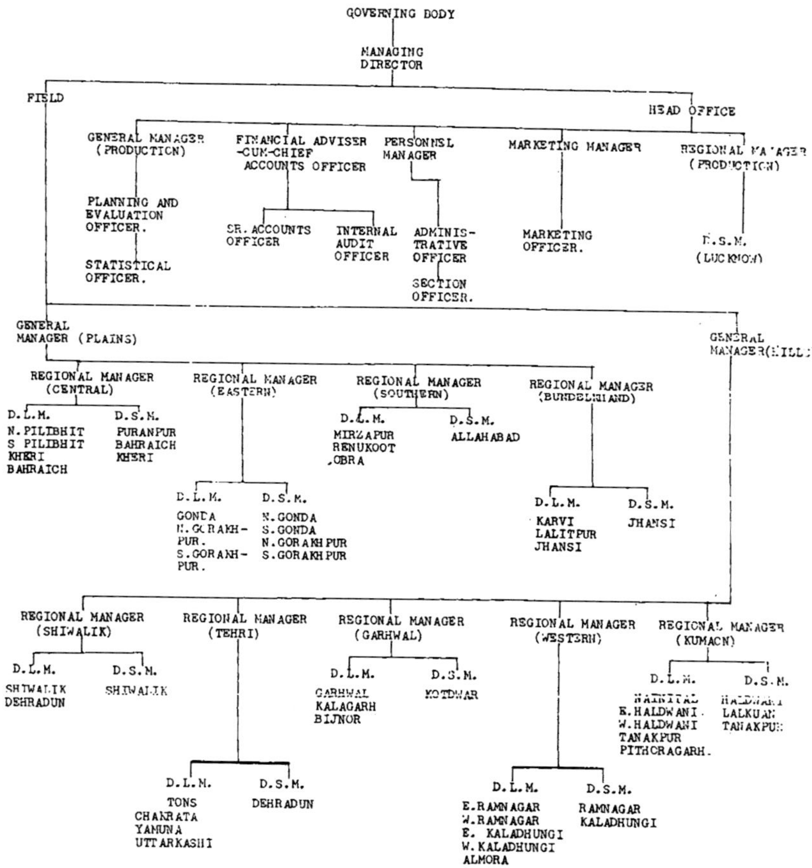

NOTE : D.L.M. = DIVISIONAL LOGGING MANAGER   D.SM. = DIVISIONAL SALES MANAGER.

E. = EAST   W. = WEST   N. = NORTH   S. = SOUTH

SOURCE - U.P. FOREST CORPORATION ANNUAL REPORT, 1984-85, ANNEXURE (I), p. 20(A).

The latest existing position of the organization as depicted in the U.P. Forest Corporation Organizational Chart 1985-86 is epitomized below:

a) The U.P. Forest Corporation functions under the direction of the Governing Body through the Managing Director.

b) The Managing Director is assisted by a number of officers at the Head Office and two General Managers in the field viz:

i) General Manager (Plains), and

ii) General Manager (Hills).

c) At the Head Office, there is General Manager (Production), Financial Adviser-cum-Chief Accounts Officer, Personnel Manager, Marketing Manager and Regional Manager (Headquarters).

i) General Manager Production is assisted by the Planning and Evaluation Officer and Statistical Officer.

ii) Financial Adviser-cum-Chief Accounts Officer is assisted by the Senior Accounts Officer and Internal Audit Officer.

iii) Personnel Manager is assisted by Administrative Officer and Section Officer.

iv) Marketing Manager is assisted by the Marketing Officer.

v) Under The Regional Manager (Headquarters) is Divisional Sales Managers (Lucknow).

d) In the field, the General Manager (Plains) is assisted by four Regional Managers of Central, Eastern, Southern and Bundelkhand regions. The General Manager (Hills) is assisted by the five Regional Managers. These Regional Managers are for Tehri, Shiwalik, Garhwal, Kumaon and Western Regions.

e) Each Regional Manager is assisted by the Divisional Logging Manager, Divisional Sales Managers and an Accounts Officer.

C H A R T - X

ORGANIZATION CHART OF U. P. FOREST CORPORATION, 1985-86.

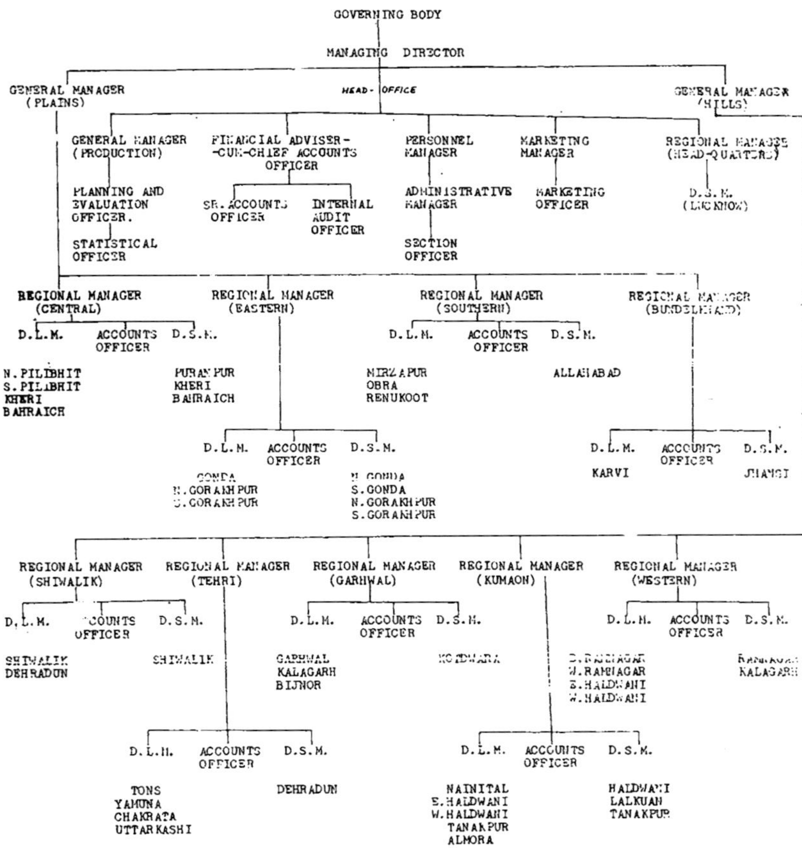

NOTE - D.L.M. = DIVISIONAL LOGGING MANAGER. D.S.M. = DIVISIONAL SALES MANAGER.

E. = EAST. W. = WEST. N. = NORTH S. = SOUTH.

SOURCE - U.P. FOREST CORPORATION, ANNUAL REPORT, 1985-86, ANNEXURE (I), p. 20 (A).

f) Regional Manager (Central) has four Divisional logging Managers at South Pilibhit, North Pilibhit, Kheri and Behraich, three Divisional Sales Managers-Puranpura, Kheri, and Baharaich and one Accounts Officer.

g) Regional Manager (Eastern) has one Accounts Officer, three Divisional Logging Managers-Gonda, North Gorakhpur, South Gorakhpur and four Divisional Sale's Managers-North Gonda, South Gonda, North Gorakhpur and South Gorakhpur.

h) The Regional Manager (Southern) is assisted by three Divisional Logging Managers-Mirzapur, Obra, Renukoot, one Divisional Sales Manager at Allahabad and one Accounts Officer.

i) Regional Managers (Bundelkhand) is assisted by the Divisional Sales Manager (Jhansi), Accounts Officer and three Divisional Logging Managers-Jhansi, Karvi, and Lalitpur.

j) The Regional Manager (Shiwalik) is assisted by the Divisional Sales Manager (Shiwalik), Accounts Officer and two Divisional Logging Managers-Shiwalik and Dehradun.

k) Regional Manager (Tehri) is assisted by Accounts Officer, Divisional Sales Manager, Dehradun and four Divisional Loogging Managers-Chakrata, Tons, Yamuna and Uttarkashi.

l) Regional Manager (Garhwal) is assisted by the Divisional Sales Manager (Kotdwara), Accounts Officer and three Divisional Logging Managers-Gerhwal, Kalagarh, Bijnor.

m) The Regional Manager (Kumaon) is assisted by the Accounts Officer, three Divisional Sales Manager -Haldwani, Lalkua, Tanakpur and five Divisional Logging Managers-Nainital, West Haldwani, East Haldwani, Tanakpur and Almora.

n) Regional Manager (Western) is assisted by Accounts Officer, two Divisional Sales Manager-Ramnagar and Kaladhungi and four Divisional Logging Managers-Ramnagar and Kaladhungi and four Divisional Logging Managers-East Ramnagar, West Ramnagar, East Kaladhungi and West Kaladhungi.

o) Under Divisional Logging Managers come Logging Officer, Logging Supervisors, Assistant Logging Supervisors, Junior Logging Assistant, Scaler and Field Assistant.

p) Under Divisional Sales Manager come Sales Officers, Logging Supervisors, Assistant Logging Supervisor, Junior Logging Assistant Scaler and Field Assistant.

Strength of the Organization

The strength of the organization, which stood oat 71 (Seventy one) in 1975 also increased. The increase in the total employment of the organization from the year 1980 to 1986 is given in the following table-

From the above table it is clear that the strength of the organization has increased to a very large extent. The number of officers which were 40 (Forty) in 1980 increased to 48 and 77, in the years 1981 and 1982 respectively. The number of officers decreased in 1983 to 75 and then again increased in 1984 and 1985 to 86 and 101 respectively. In 1986, the number of officers further increased to 109.

The number of class three employees also showed a gradual increase. The number of third class employees which stood at 296 in 1980 increased to 306 in 1981, 391 in 1982, 556 in 1983, 584 in 1984, 786 in 1985. The number of third class employees further increased to 837 in 1986. Similarly, the strength of fourth class employees has increased from 390 to 400, 476, 609, 798 and 1102 in the years 1980, 1981, 1982, 1983, 1984, and 1985 respectively. The number of fourth

	Years						
Category	*1980*	*1981*	*1982*	*1983*	*1984*	*1985*	*1986*
Officers	40	48	77	75	86	101	109
Third	296	306	391	556	584	786	837
Fourth	390	400	476	609	798	1102	1370
Total	726	750	944	1240	1468	1989	2316

Source : Own Compilation on the basis of Annual Report, U.P. Forest Corporation.

class employees also increased in 1986 and became 1370. This increase in the stregnth of the organisation has taken place due to increase in decentralisation of the organisation. This has helped in increasing the level of specialisation and assisted in better management of the U.P. Forest Corporation.

Besides this regular employment, large employment was also created for the labour force. The total labourers employed during 1980-86 can be shown with the help of table given below:

From the table above, it is evident that the Corporation has provided large employment also to the labour force. The number of labour employed per day constantly increased between 1980 to 1986. In 1979-80 only 6,000 labours were employed per day. The increased to 7,000 labours per day in 1980-81 and 3,000 labours per day in 1981-82. In 1982-83, the number of labour employed per day increased to tremendously and became 12,000. The labour employed per day kept on increasisng and became 13,900 in 1983-84, 14,000 in 1984-85 and 14,200 in 1985-86.

The U.P. Forest Corporation very frequently changed its organizational structure in order to meet the increasing responsibilities entrusted to it. The changes in the organisational structure have

	Years						
	1979-80	*1980-81*	*1981-82*	*1982-83*	*1983-84*	*1984-85*	*1985-86*
Labour employed per day (in thousand)	6.0	7.0	8.0	12.0	13.9	14.0	14.2

Source: Own compilation on the basis of Annual Report,U.P. Forest Corporation.

been beneficial to the Corporation in bringing specialization. The various beneficial effects of changing organizational pattern are as follows:

(a) The Corporation has brought specialization in its management by bringing decentralization in its management into two branches:

viz. Head Office and Field. The Corporation has further decentralized its organisation in the field operations into two specialised branches, namely; logging and sales. Such a decentralization has helped in improving the efficiency and also in bringing division of labour.

(b) The expansion of the organisation has helped in increasing the job opportunities in the forestry sector thereby increasing the level of employment.

(c) The Corporation is employing trained personnel in various activities. This has helped in giving employment to educated unemployed youths who are trained in different branches. The use of trained and efficient labour has helped in improving the efficiency and is also helpful in better utilization on resources.

(d) The Corporation is also providing training to its officials so that wastage may be reduced at all levels and more economic utilization may be done.

(e) The Orgnization of the Corporation is having all the specialized braches, viz. Marketing, Logging, Accounts, Administrative, Planning and Statistics.

Though, the Corporation is having specialized organisation, yet it suffers of certain shortcomings as follows:

i) Most of the post of the Senior Officers of the Corporation are filled by the Forest Department Officers, who are on deputation to the Corporation. These Officers are not suited for performing the marketing activity of the Corporation, which has a business approach. Hence, the Corporation should go in for its own direct employment. The Corporation should employ person, who are having business aptitude.

ii) The Corporation is lacking in its own Research and Training wing. The Corporation should have its own training wing, so that it may not have to face problems in providing technical training to its employees.

iii) The Corporation should have a Forest Economist to evaluate the working of the Corporation from time to time. This will help in formulation of the policy of the Corporation and will also help in the development of the Corporation on desired lines.

iv) The Corporation has failed to provide permanent employment to the labour and has also failed in the development of labour co-operatives.

v) The Corporation has no Advisory Committee to study the working of the Corporation and give suitable suggestion for proper implementation of various programme.

Though, the organization suffers with the above mentioned shortcomings, yet is working very efficiently.

References

1. A.R. Maslekar, "Managing the Forests", 1983, p. 37.
2. P.N. Khandawala, "Design of Organization; Harcourt Brace Jovanovick International, New York, 1977.
3. A.R. Maslekar, "Managing the Forests", 1983, pp. 37-38.
4. D.R. Johnston, A.J. Grayson and R.T. Bradley: "Forest Planning", Faber & Faber Ltd., 1967.
5. L.C. Sharma, "Forest Economics, Planning & Management, pp. 278-28, Dehradun, 1980.
6. A.R. Maslekar, "Managing the Forests", 1983, p. 38.
7. U.P. Forest Corporation, A Short Note on its Activities, 1974-75 to 1984-85, p. 6.

CHAPTER SIX

Processing SystemUnder U.P. Forest Corporation

Forestry operations are closely linked with scientific management of the forests. The major part of scientific management involves the study of forest utilization. In the beginning, when population was not so much, the extraction was confined to only few species which were useful for construction of dwellings, agriculture implements, etc. But with the increase in human population and large-scale disappearance of forests, the value of forest products increased and consequently the attitude towards forests also changed. Restrictions were placed on the extraction and clearance of forest produce. Upto this stage, the utilisers of forest produce were free to adopt their own method of felling and extraction.

Towards the middle of the twentieth century, lot of restrictions were imposed on felling and extraction. New methods of felling, conversion and transportation were evolved to bring more scientific methods of working and management into practice, to bring optimum utilization of forest resources.

Forest utilization is defined as the process of harvesting converting and disposing of forest produce and other resources of forests.[1] It deals with felling of trees, their extraction, processing and transporting the forest produce to the market for the purpose of sale i.e. for finally marketing the forest produce to the consumers. Extraction, felling, conversion and transportation have principally to do with the production of logs which are the chief primary products of the forests. It also includes the production of large quantities of scantlings, sleepers, poles, pilings, and other miscellaneous forms. Practically all is produced as logs in round form of varying lengths, which are later converted into forms of the final product, generally square or rectangular shapes.

Forest utilization constitutes major part of forestry operation in the country. To achieve complete and efficient utilization of forest produce i.e. standing timber, it is necessary to have modern logging techniques of harvesting timber, scientific methods of collection of Minor Forest Produce, attention to wood preservation and seasoning, saw-milling and establishment of many forest based industries. All this reflects the trend of forest utilization[2].

The problem of forest utilization has greatly intensified because of great distances between the centres of production and of consumption, which increases the cost of transportation. The cost of felling, conversion and transportation forms a considerable part of the cost of converting timber into variable form. The cost of felling, conversion and transport is approximately 50 per cent of the total cost of delivery of timber in the market. The tree prices are lower as compared to the value of finished products, they generally represent 10 per cent to 22 per cent of the value of saleable products. Great wastage has attended the conversion of trees into usable forms. A large proportion of timber of trees felled in our forest is still wasted. Extensive wastage may occur in the forests because profitable means of utilization may be limited. The wastage in the hilly regions of the forests is comparatively more than that of plain. But even in plains considerable percentage of volume of timber is left to rot. Large quantities of material are left in the forest because of broken tops, broken logs or because of disease and insects, rendering their condition unfit for utilization. There is also heavy wastage in the process of conversion of raw-material in the form of saw dust and off cuts and pharras slates; which are seldom usable except as firewood.

The extent of utilization is now becoming directly related to market prices. Lower price means less of utilization and increase in wastage, while higher price means low wastage and more complete and efficient utilization

A study was conducted in Himalayas which gave the approximate wastage taking place as follows :

Table - 1

WASTAGE OF TIMBER.

Sl No	*Items of work*	*Approx. Percentage of wastage of timber*
I.	Felling High stump, splitting, felling in wrong direction, destruction in felling.	9% to 14%
II.	SAWING-	
	a) Hand sawing	27% to 32%
	b) Pulpwood	4% to 5%
III.	MINOR TRANSPORTATION -	
	a) Chutes and dry slides	1%
	b) Wet slides including wastage of timber in construction of slides.	2% to 3%
	c) Telescopic floating in small Nallas and abnormal conditions.	2% to 3%
IV.	MAJOR TRANSPORTATION -	
	a) Normal river loss (without floods).	1% to 3%
	b) Theft of timber enroute.	2%
	Total wastage	48% to 63%

SOURCE: Tribhuwan Mehta, "A Handbook of Forest Utilization," 1981, p.6.

The total volume of usable rough timber from the average tree is therefore, only 37 per cent to 52 per cent, depending on species, region and local conditions. The greatest wastage in felling and conversion of our forests is found in the Himalayas forests, particularly in such species as fir, deodar and spruce. The wastage in conversion of teak, sal and some of the valuable hardwoods is generally less.

The wastage is generally attributed to the following reasons:

(i) Carelessness in cutting high stumps. Stumps as high as 30 - 45 cm are left, which results in considerable wastage of the most valuable portion of the tree.

(ii) Small and short top logs are frequently left in the forests because these can not be handled profitably.

(iii) Use of axe in felling trees results in considerable wastage.

(iv) Improper measuring of lengths.

(v) The practice of cutting logs in even lengths instead of odd and even lengths.

(vi) Breakage in felling, due to carelessness, resulting in shattered trunks, broken tops, split logs, etc. The damage in hills is due to rough terrain. It is also due to trees felling on other felled trees or wind falls or across nullahs and depressions.

(vii) Losses in transportation- Some timber is often lost in floating operations; it is either stolen or is used as fire-wood by the labourers. The timber is left along stream banks as waif or lost as sinkers in floating and rafting. It is also shattered due to rocky nature of stream beds and in jams.

(viii) Losses due to insect and fungal damage.

(ix) Damage to the remaining crop in the forest on account of lack of attention to felling, rolling or using cable-way hauling system.

(x) Leaving large quantities of material behind in the forests which could be profitably converted into small sizes or utilized as charcoal and or firewood.

There has been great wastage taking place in forest. Every possible step is being taken to avoid wastage of forest produce and to maintain perpetuation of forests, continuity of production and sustained yield. The method by which mature forests are utilized are an important factor to achieve the above basis of sustained yield. Better utilization practices have to be achieved and compatible with the silvicultural practices to achieve complete utilization of forest products. The cost of production is to be decreased, increasing demand for forest product is to be met and determination of profitable outlets of timber and reduction of wastage are the basis of economic utilization.

The U.P. Forest Department realised that the contractors did not deal fairly with the forest labour and also with the forest produce. So, the State Government decided to eliminate the middlemen and contractors (former pattern of working) working system from forest. In place of it, the Government-undertook the responsibility of working through U.P. Forest Corporation. The Forest Corporation was entrusted with the full responsibility to perform the complete forestry operation to bring efficient utilization of the forest produce

The U.P. Forest Corporation envisaged certain improvement in working through better harvesting practices, fair and higher wages to labour and better amenities of working and so on. The corporation used more and more modern and scientific methods of timber harvesting and logging techniques for the processing of forest produce. The Corporation specialised every step of the forestry operation. The Forest Corporation divided the complete forestry operation into two broad branches, viz:

i) Processing of Forest Produce.

ii) Marketing of Forest Produce.

Of these two branches, 'Processing' of forest, produce is dealt with in this chapter and Marketing in the next. The forestry operations includes felling, extraction, conversion, transportation, storing and marketing of forest produce. Processing of forest produce include felling, extraction, conversion and transportation of forest produce while Marketing includes storing, grading and marketing of forest produce.

The U.P. Forest Corporation has been handed over the specialised function of exploitation of forest resources. The exploitation of resources involves the entire process from marketing by forest department and then handing over of the lots with sale lists to the Corporation) to transportation. A detailed study of processing of forest produce by the U.P. Forest Corporation is done under the following headings, viz:

(1) Marking of Forest Produce and Preparation of Sale-List.

(2) Transfer of marked coupes to Corporation.

(3) Felling.

(4) Conversion - Billeting, Debarking, Logging, etc.

(5) Recording.

(6) Transportation.

(7) Transit Depots.

(8) Fixation of Royalty.

(9) Labour/Organization.

(10) Progress of Extraction.

Marking of Forest Produce

The marking of forest produce i.e. trees, is done by the Forest Department strictly in accordance to the prescriptions laid down in the working plans of the respective Forest Divisions. The marking was based on the three systems of feelings, namely:

i) The Uniform System.

ii) The Copice System.

iii) The Selection System.

In the uniform system, the trees of uniform age are marked for felling so as to get progressively sustained yield.

In Copice system, the entire crop is cleàrfelled at one time. In selection system, the felling is carried out in the whole area with trees of uneven age or mixed age of exploitable diameter being marked for the purpose of felling.

Of these three systems, the U.P. Forest Department generally uses the selection system of felling. The Forest Department marks the following types of trees:

a) The trees which are mentioned in the working plan prescriptions or which are required for some special schemes of the Government.

b) All the trees which are dead or are dying or are diseased are marked for felling.

c) Old, mature and overmature trees are marked.

d) Trees which are dangerous to the surroundings are also marked.

e) Sometimes, trees are also marked under special schemes, like dams, etc. or to meet the special demands.

The Forest Department marks the trees at two places, namely:

i) At 4.5" above the ground.

ii) At breast height level, that is, 4' 6" above the ground level.

The size of the mark (also known as stump mark or 'gadava') is 9" x 9".

After completing the marking, the marking record is compiled by the Forest Rangers, classifying the marked trees species wise in 4" (10 cm.) diameter classes and grouping them into sound, fit and unfit categories. After properly scrutinizing and grouping the trees into 10 cm diameter classes, an abstract of such enumeration is prepared, forming lots of suitable sizes. List of lots is then submitted by all the Forest Rangers of a division to the DIvisional Forest Officer then prepares a consolidated sale list.

Transfer of Marked Coupe to Corporation[3]--The consolidated sale-list of lots is then handed over to the Forest Corporation. The handing over and taking over of lots in which the Forest Corporation has to work is done in writing. For this, ordinarily the Chief Logging Officer will inform the Divisional Forest Officer and the Range Officer in writing and will submit a written information of taking over of charge before starting the work. In respect to those lots in which work is to be done by the Forest Corporation, the procedure for taking overcharge will be same as applicable to the contractors. The authorized representative of the Forest Corporation will sign the register, the map (in which lot boundaries are shown) and other records and from this very date the Forest Corporation will be responsible for these lots. The consolidated sale list of lots is handed over to the Corporation by 31st July. In the case of annual auctions, the handing over of lots takes place latest by 31st October and the lots of subsidiary auction and cultural operation will be handed over latest by 31st January. In the lots so handed over, the Corporation will have to complete work within the time limit ordinarily to be allowed to contractors in the near of allotment.

After receiving the lots, the Forest Corporation checks the lot in accordance with the sale list. If there is any appreciable difference, the Forest Corporation can get the recounting of trees done in any lot. The rules applied to the Forest Corporation for recounting will be the same as prescribed under the sale-rules for the contractors.

Felling

After completing the final handing over of lots,the Forest Corporation starts the felling operation. The felling operation is carried out by the Forest Corporation with the help of trained labour and logging experts or through the Forest labour Co-operatives. The U.P. Forest Corporation does felling strictly in accordance with the silviculture norms. The following felling rules are observed by the Forest Corporation in order to maximize production of sound timber and to avoid damage to the surrounding vegetation of the forests.

(a) Felling of trees is done as near the ground as possible. In the areas where clear felling is done, the felling of trees is done flush with ground and in other areas where selection felling is done, the trees are felled just above the lower mark. This helps in reducing wastage of timber.

(b) The felling is done in such a manner and in such a direction in which least damage is caused to the tree and to the surrounding forest. For example, trees should not be allowed to fall across boulders or spurs or sharp depression or on other lying trees. They should be felled in uphill direction as this will have the smallest angle for them to fall. This will help the trees to fall parallel to the ground. Trees on hill side have a tendency to lean outwards are generally more branchy on the lower side, and thereby have the tendency to fall that way. Such trees are first lopped and the fall directed by the use of rope or the cable puller.

(c) Felling is done in vertical strips and the trees felled in herring bone fashion. In this way all trees that start sliding or rolling are held up by the standing trees.

(d) Felling along contour in undulating ground is avoided as it results in the trees breaking their backs and causing great damage to the timber.

(e) Sometimes, it is more advantageous to fell trees down hill, in which case no lopping should be done as branches act as a brake. This type of felling is resorted only when there is no other alternative possible.

(f) Trees should be felled in such a place where it is possible to convert and extract timber. It is always better not to fell such trees, similarly, it is not worth-while felling those trees which are likely to be smashed to pieces in falling.

(g) Trees should not be felled during a strong wind, when it is not possible to ensure the fall of the tree in the desired direction and the work itself is exceeding difficult and dangerous.

(h) Felling is begun at the top of the slope and proceeds in a down hill direction. In such situation, if any tree starts sliding down, then such trees will be held up by the stems below as the lower part of the forest has not been worked. Such felling will be advantageous as it will not cause any damage to the previously felled trees and of stacks of timber.

i) Lastly, felling should be concentrated as much as possible and the felling area should be divided into sections so that the work is carried out in a controlled manner and is under strict supervision.

Felling is an art and is done with the help of skilled labour, which is having technical knowledge. Before starting felling, first the place of work is prepared. That is, all weeds, shrubs and branches which would hinder work are first removed with the help of 'bill-hook'. In order to saw the stump deep near the ground, leaves, earth and stones are removed with pick-axe. The stems with thick bark should be debarked upto shoulder height and the butt end of the stems is rounded off. After this, the felling operation starts:

Felling is doen in four ways [4] :

i) Felling with axe alone

ii) Felling with axe and saw.

iii) Felling with saw alone.

iv) Felling by roots.

Felling with Axe

Felling with axe, a notch or undercut is chopped on the trimmed stem, in the direction of felling. The notch is cut about 2/3rd of the diameter of the tree (3/4rd diameter in case of big trees) so that the centre of gravity of the tree is cut by this initial cut. The notch has a level base and a straight inner line called tipping edge. This notch makes it possible for the tree to fall over tipping edge and break it from stump. The cut is chopped flat and horizontal with the help of axe. Then the next cut is made exactly opposite to this cut, about 10 - 15 cm. above. Both the cuts are parallel. To have a desired fall, ropes are also used. The figure given below shows felling with axe:

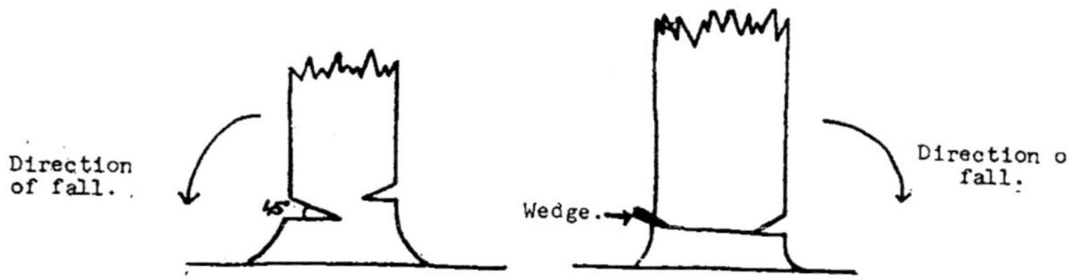

The Forest Corporation very rarely uses this method as this method causes large wastage of timber and other damage.

Felling with Axe and Saw

On the trimmed stem, the notch or undercut is made on the side of the direction of fall with the help of axe. The notch should be flat and not too big. This notch makes the tree fall possible. Then the saw cut is made by cross-cut, saw, about 2.5 cm above the notch on exactly the opposite side of notch. As soon as, saw has penetrated about twice the breadth of saw blade, a wedge is inserted and sawing continued. The wedges help in manipulating the direction of fall. The wedge and saw work is done carefully and without any haste. In this way, when cracking starts, the saw is left in the cut or pulled out before the tree begins to fall. Now the tree starts falling gradually. Sometimes ropes or cables are also used to direct the fall.

Felling with axe and saw is shown below :

This method is being frequently used by the Forest Corporation.

Felling with Saw alone

In this, the felling is completed only with a saw. The saw cut is made and the tree is tipped by continuous driving in of wedges. In case of power saws, an undercut is made with the help of saw itself, this corresponds to the undercut which was made with the help of an axe in felling the tree with axe and saw combined. This method helps in saving 3 per cent to 7 per cent of the best timber of the butt end, the cut being essentially made in the stump rather than the tree.

Felling with saw alone is shown below:

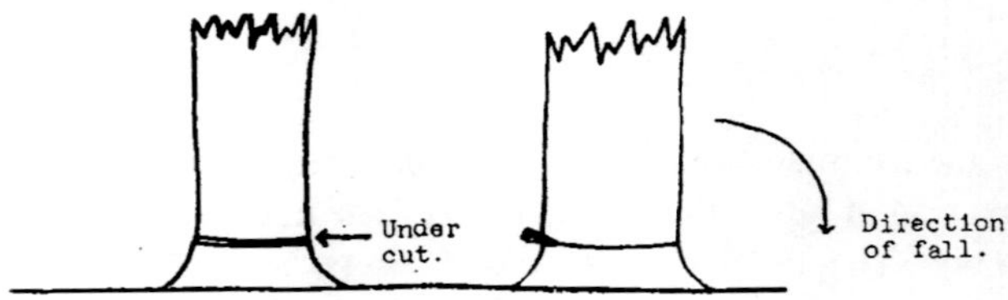

The direction of felling is also manipulated by the help of ropes or cables. The direction is also manipulated with the help of wedges.

Felling by Roots

Felling by roots is also practiced where the ground is needed for plantation or agriculture purposes or for road construction or when the very valuable tree is involved having the valuable underground portion. Felling by roots help in complete utilization. In this method, the earth is dug away from roots and then these exposed roots are severed with an axe. The fall of the tree is then accelerated by the use of ropes, tree winches or a cable puller. Many mechanised devices like Bull-Dozers, etc. can be used for uprooting the tree. But these devices are not very popular in the U.P. Forest, except where clearfelling is done for plantation.

Of these four methods, the most prominently used by the Forest Corporation are namely:

i) Felling with saw alone.

ii) Felling with axe and saw.

The U.P. Corporation starts the field operation i.e. felling in forest generally in October and ends it by June. During the rainy season, the extraction of already felled forest produce is done. In the plains the working season is mostly confined to winter i.e. from October to March and in the Hills, mostly felling takes place between March and June. The Corporation never undertake the felling operation during the summer season because there being every possibility of splitting, cracking and rapid drying of log taking place.

The Forest produce which is allotted by the forest department to the Forest Corporation for exploitation is hammered separately by the Forest Corporation. For this, the Forest Corporation will get their hammers registered and each registered hammer will be used in one lot/unit only.

After the felling is over, the joint inspection of Unit Officer of the Forest Corporation and the representative of the Forest Department takes place. In the joint inspection, the Unit Officer of the Forest Corporation will have his own personal hammer for marking the illicit felled and damaged produce. If there is any dispute between the Officials of Forest Department and Forest Corporation over any kind of damage or loss, it will be subjected to decision by the Divisional Forest Officer. If the Forest Corporation is not satisfied with the decision of the Divisional Forest Officer, then it will be put before the concerned Conservator of Forests, whose decision will be final and binding on all.

In the lots being worked by the Forest Corporation, the number of stump counters will ordinarily be 50% less than the number of stump counters engaged in the lots being worked by the contractors. These stump counters properly maintain record of the size and quantity of timber being extracted from the felled trees as well as of the damage to forest produce and its nature, if any, i.e. avoidable or unavoidable.

Conversion

Generally, the logs are of long lengths of both, even and odd sizes. The conversion of bole into logs is done mostly with cross-cut saws. The cuts are executed at right angles to the axis of the stem.

The conversion of bole into logs is done preferably at site because of following important reasons:

i) Conversion assists in reducing the weight of the tree.

ii) Helps in eliminating all defects and unsaleable proportion of tree.

iii) Lastly, conversion helps in using the modern and mechanised methods of transportation.

The dictum, Necessity is the mother of invention, is aptly applicable to the logging operation done by Forest Corporation. The Forest Corporation, since it started working, has developed the process of felling and conversion. The Corporation has put an end to the use of axe and started using saws of modern and improved nature. It has since 1980-81 started using 'Pit Sawing Method' for feeling trees, which has increased the production to 70% to 75%, instead of 50% which was the case in contractors system.

After felling the free 10 the ground, the conversion of tree into logs is done. The conversion of tree involves trimming of the branches and sawing the clean bole and the larger branches into lengths of required sizes. Then the tree is converted into logs. In the whole process, Forest Corporation works on the principle of least possible wastage. The conversion by the Forest Corporation is done in the following manners.

a) First, all the branches of the tree are trimmed. The larger branches of the tree are converted into lengths of required sizes, depending upon the type of branch. For this purpose, cross-cutting sawa are used.

b) Secondly, fashioning is done. This includes leveling bigger knots, flutes and buttresses and other projections of the felled tree and converting the tree into bole. For this purpose, axe and saws, both are used.

c) Lastly, after completing the fashioning the bole is converted into logs by cross cutting. The size of the log depends upon the market requirements and the methods of transportation.

Logging being a technical work, the Forest Corporation has taken the responsibility of providing basic logging training to the employees. The Corporation has also employed technically sound officials like logging supervisors, logging assistants and logging officers to improve the level of logging operation and to increase the efficiency of labour and to reduce the damage of forest produce. The Corporation has also adopted the use of 'Gravity ropeways' and 'Skyline-Cranes' in hills for off-road transportation to avoid destruction of crop and erosion of soil.

Recording

When the conversion of bole into logs is over, the recording of the produce is done in a systematic manner in the Register. The forest produce is properly measured, graded and numbered by the logging staff, the Scalers and the Field Assistant and then recorded in the register. The following entries of the register are completed, namely-the measurement (length) of the logs, girth of logs, grade of logs and the number of logs.

After completing the register the passing of produce takes place. For passing the logging supervisor issues a permit known as 'Rawanna'. The total produce which is passed is noted down in a proforma with the complete measurement of length and girth, grade and number, and then transported. Two copies of the proforma having the details of produce, which is being transported is sent with the produce for the purpose of checking at check-post and finally at the sale-depot (immediately before entering the depot and after unloading). After the checking of produce is over at the depot, one copy is sent back with the necessary comments relating the produce.

Transportation

Timber extraction is pre-eminently a problem of transportation and moving bulky and heavy product of Forests, logs and converted timber from the forests to the saw mill for further conversion or to the ultimate market. Transportation constitutes almost 60% to 70% of the total cost of delivery of product to the market and that is why utmost care is taken in choosing the mode of transportation.

Transport of timber from the area of exploitation to the final destruction, i.e. the sale depot or other consuming centres is divided into two parts, namely:

i) Minor Transportation as it is also called 'Off-Road Transportation'. Minor transportation consists of assembling of logs and converted timber together at a common point with in the forest from where eventually the transportation to the final destination can take place.

ii) Major Transportation deals with final transportation of forest produce from the road-side, transit depots, etc. to sale depots.

The chief methods of transportation used by the Utter Pradesh Forest Corporation are as follows:

(1) *Carriage by Men* - Generally, small and medium sized timber like sleepers, poles and fuel wood are carried by men to the road side on their shoulders or back.

(2) *Dragging by Animals* - The Forest Corporation also uses animals for the purpose of dragging timber to the road-side. For this purpose mules and ponies are used.

(3) *Carriage by Carts*- Corporation uses this method very extensively to transport timber or forest produce from the felling coupe to the transit depots or to the road-side.

(4) *Sliding* - Forest Corporation uses sliding as one of the method of dragging timber to the road-side. For this purpose results in damage to logs and the forest.

(5) *Floating* - Forest Corporation uses floating of timber which has no tendency to sink is floated. This method is largely used in the hilly regions.

(6) *Rafting* - Rafting constitutes a part of floating. In rafting, a quantity of timber is firmly bound together into a number of sections known as a raft. Rafting is used for the delivery of log or sawn

timber in the depots situated on the river sides. Rafting is done generally in winter months, but in no case during the rainy seasons.

(7) *Booms* - Boombs means a floating obstruction or a barricade across river to intercept the floating timber coming down the river, with a view to catch, collect and diverting them into other channels or to a place where rafts are made after tying the free floating timber. The Corporation pays the Forest Department, the charges of use of booms.

(8) *Skyline-Cranes* - The Forest Corporation is also using skyline-crances for transportation of logs of different weights (not above the prescribed weight which depends upon allowable tension) over an aerial cableway on steep hill slopes where construction of road is not feasible or economical.

(9) *Motor transportation* - The main method adopted for transporting timber is through motor transport. Motor transportation is very widely used by the Forest Corporation for both minor transportation and major transportation purposes. Trucks and Tractors are used for carrying timber from the forest coupes to the sale-depots, while truck alone are used for finally transporting timber from the roadsides, transit depots, booms, or common places of collection to the final destruction i.e. the Corporation Sale-Depots. Road transportation is done by the Forest Corporation trucks and by trucks on hire from private parties. The Corporation employs the trucks on the rate contact basis or through tenders, for a specific period of time on certain terms and conditions. The Forest Corporation has also to pay the road fees, etc. under the existing rules, like other contractors.

It is very important to note that in the process of collecting of timber and transportation, the skilled manual labour is employed, by the Forest Corporation. The Forest Corporation has not so far been able to use modern machines for the purpose of loading and unloading of forest produce. Due to this reason lot of time is consumed in performing these operations and also amounts to large amount of wastage of energy of labour. The Corporation can take effective measures and introduce power lifters and cranes for this purpose.

In the final stages of transportation utmost care is taken throughout the whole time of transportation from the site of conversion i.e. the forest coupe to the sale-depot. This is done by a careful system of timber accountancy, proper scrutiny and checking or roads and check posts. This will help in avoiding loss of timber in transportation and will help the Corporation in reaching the complete forest produce to the Corporation Sale Depots. The checking on the way is done by both, the officials of Forest Corporation and those of Forest Department.

Transit Depots

Transit depots are the intermediate depot inside or outside the forest, where forest produce is temporarily stored, pending its onward dispatch to sale-depot. Transit depots by the Forest Corporation are used mainly for collecting, checking and examining the timber, where transportation is done by land and for catching, collecting, examining and rafting of water borne timbers. Transit depots are temporarily made near roadsides, launching depots, Ropeways heads and boom stations.

All timber that is exploited in the interior of forest is first brought to the transit depot with the help of minor transportation. At the transit depot it is checked and stacked. Proper marking numbering and recording of the produce is done in the transit depot. It is finally tansported to the Corporation sale-depot. Two copies of the proforma containing the details of the produce transported is also sent alongwith the produce, in order to facilitate early checking of produce at the check post and sale depot. To avoid any inconvenience on the way, the Corporation issues a permit known as 'Rawanna' (which is registered) of the forest produce transported.

The Corporation will also pay the forest department lease rent at the rates specified by it for each and every depot established by it (the Corporation) and will follow all rules for the upkeep of depots.

After completing the work in the forest coupe, the Forest Corporation will give in writing to Forest Department, the intimation of the resignation of lot in which the work has been completed. The resignation by the Forest Corporation will be accepted within one month by the Forest Department after joint inspection.

Fixation Of Royalty AndPayment

The determination of royalty payable by the Forest Corporation will be made according to the G.O. No. 11578/14-1-78-1(7)/76 dated 5th December 1978 and Government D.O. No. 10758/14-1-1977 dated 26th September 1977.

The following is the basis of calculation of royalty :

(a) Where the Forest Corporation is working in the entire Forest Division, the figures of royalty payable by the Forest Corporation will be worked out after calculating the simple average of the weighed average royalty figures for each of the previous three years, obtained in open auctions.

(b) Where the Forest Corporation is not working in the entire Forest Division, the royalty payable by the Forest Corporation may be calculated separately for those ranges where the Forest Corporation is working. The royalty payable to the Forest Department may be calculated after working out simple average of last three years aucions royalty for each range (the auction royalty for each year will be the weighed average for the whole range).

(c) Where the Forest Corporation is working partly in a range there is no other way out except that the royalty payable by the Forest Corporation may be calculated on the basis of the royalty obtained in auctions in the past three years in comparable lots.

(d) Where the Forest Corporation works the soft wood trees for the supply of material to the industrial units the royalty payable by the Forest Corporation will be on the basis of the cost of soft wood.

The figures of royalty obtained in the auctions during past three years will not be available in respect of those Forest Divisions/Ranges where the Forest Corporation has been working for some time past. There is no other choice except that the figures for each year in the past three years in respect of the adjoining Forest Division/Range where the auction might have taken place may be worked out for the Forest Division/Range where the Forest Corporation is working and on the basis of its trend, the 'Simple average' of the figure may be worked out which will give the royalty payable by the Forest Corporation.

An example of the calculation of royalty is given in the Appendix-III. The example clarifies the principle given above.

The Government further ordered that the Forest Corporation will also give half of its net annual project in addition to royalty. But later, vide G.O. No. 11578/14-1-78-1(7)/76, dated 5th December, 1978, the provision of payment of half of its net annual profit was done away with. These orders were made effective from the date of previous orders, that is, February 18, 1976.

In these orders, the second sentence of the previous order was modified in that the royalty of the allotted lots to be charged from the Forest Corporation will not be less than the average of the prices obtained in the last three years public auction of the comparable lots. And a new clause was added that the price of every lot in a year will be determined after taking into consideration all the relevant factors.

It appears that the Government gave every possible latitude to the Forest Corporation in determining the royalty payable by it and tried to save it from any possible losses. This sort of lenient attitude by the Government was perhaps due to the fact that for some years in the beginning the Forest Secretary to the U.P. Government was ex-officio-Chairman of the U.P. Forest Corporation. So, it was but natural that the Forest Corporation had so much advantage.

The payment of royalty by the Forest Corporation will be on instalment basis. The following will be the procedure of payment of instalment of royalty.

(I) Except the lots of conifers, for the lots of other species, the dates of installments for the Corporation will be determined in the following manners:

a) One-third price of the lot on March 1- in the year of the lot.

b) Second 1/3rd price of the lot-on June 1-in the year of lot.

c) Remaining price of the lot-on Sept. 1-in the year of the lot.

(2) For lots of conifers, the instalments would be as under:

A) *For Chir lots*

i) For lots which are allotted in Sept.-October-

a) If 'Nikasi' be done by road, then the dates of instalments will be according to the above rule 10.

b) If 'Nikasi' be done by floating the final date of payment will be about 17 months after on March 1 in the third year. For example-the final date for Chir lots allotted in Sept. 1976 will be March 1, 1978.

ii) It some lots are allotted after September/October then the Divisional Forest Officer will extend the dates of instalments of such lots by the same period by which the allotment of the lots may have been delayed.

B) For Deodar, Kail and Fir lots:

1) For lots being allotted in December-January :-

(a)For lots, 'Nikasi' of which is to be done by road:

i) One third price on the first of next September.

ii) Second 1/3rd price on the next first June.

iii) Third instalment on the next first December.

For example-the instalments of lots allotted in December 1977/ January 1978 will be as under:

- First instalment -- 1.9.78.
- Second instalment -- 1.6.79.
- Third instalment -- 1.12.79.

b) If 'Nikasi' is to be done by floating the final date of payment will be about 26 months after, on 1st April in the third year.

For example-for the lots allotted in December 1977/January 1978, the date of final payment will be April 1, 1980.

ii) If some lots are allotted after December/January, then the Divisional Forest Officer will extend the dates of instalments of such lots by the same period by which the allotment of lots may have been done after December/January.

C) If the 'Nikasi' in some lots of Chir, Deodar, etc. be done partly by ropeway, and partly by road, then the due dates of instalments will be the same as for the lots in which 'Nikasi' is done wholly by floating.

(3) In case of non-payment of instalment on the above stated prescribed dates, late fee will be charged from the Forest Corporation under the ordinary rules of the Circle at the same rate as is charged from the contractors.

The total Royalty paid by the Foreset Corporation from 1979-80 onwards in shown in the following table:

Years	*Royalty paid by Corporation to Forest Department (in lac Rupees)*
1974-75	25.50
1979-80	656.65
1980-81	1043.91
1981-82	1865.18
1982-83	3368.19
1983-84	4539.96
1984-85	3995.41
1985-86	4932.20
1986-87	4950.00

Source : Own compilation based on Annual Reports, U.P. Forest Corporation

From the above table it is very clear that the royalty paid in the first year was very less i.e. Rs. 25.50 lacs, because the Corporation had just started working. But from 1979-80, the royalty paid by the Corporation started increasing because the Forest Corporation field of operation increased every year. The royalty paid by Corporation increased from Rs. 656.65 lacs in 1979-80 to Rs. 1043.91 lacs in 1980-81. In 1981-82, more areas were transferred to the Corporation which caused an increase in royalty to Rs. 1865-18 lacs. By the year 1982-83, Corporation had started working the entire timber of the State and due to this the royalty paid by the Corporation also increased to Rs. 3368.19 lacs. In 1983-84, the royalty paid by Corporation increased to Rs. 4539.96 lacs but decreased in 1984-85 to Rs. 3995.41 lacs. And it is estimated that the Corporation will pay. Rs. 4932.20 lacs in 1985-86 as royalty and Rs. 4950.00 lacs in 1986-87 (as budgeted). It is important to note that the royalty did not increase in the same proportion as is the increase in the price index.

Besides the above mentioned conditions, the Forest Corporation will also follow the following conditions and rules:

1) The following procedure will be followed in the event of non-completion of work in the allotted lots by the Corporation within the prescribed period of working:-

   a) The Forest Corporation will apply for extension, which will be accepted under the existing rules.

   b) In a lot in which the Corporation does not start work at all, the rate of royalty of that year will be applied in which the Corporation does work.

   c) For the lots in which the work remains incomplete extension fee will be payable by the Corporation for extension next year and late fees will also be payable under ordinary rules on the royalty not paid on the specified dates. Their rates will be the same as are applicable to contractors.

2) The other terms and conditions of sale, which may ordinarily be prevalent in the concerned forest division, will apply to the Forest Corporation in the same manner as are applicable to contractors.

3) Ordinarily, due to non-compliance of rules or delay in the payment, neither any restriction will be imposed on the extraction of produce from the lots of the Corporation nor any such restriction will be imposed on the sale of forest produce.

If gross violation of any rules, terms and conditions is committed by the Corporation, this fact will immediately be brought to the notice of all authorities.

4) If the need is felt for any amendment in the policy of royalty fixation (determination), the committee constituted as follows will submit its recommendations to the Chief Conservator of Forests and the Government:

   a) Addl. Chief Conservator of Forests (Management), U.P.

   b) Managing Director, U.P. Forest Corporation.

   c) Concerned Conservators of Forests.

5) As far as possible, the Forest Department will not hold auction of its lots on the dates fixed by the Forest Corporation for their auctions. In this connection the Forest Corporation will intimate the dates of its auction to all the concerned divisional forest officers and others. If the Corporation wishes to hold auction on dates other than these, it will seek prior written sanction (concurrence) of the Addl. C.C.F. (Management).

6) If necessity arises to make any modification/amendment or to add any additional rule to the aforesaid rules, then amendment in the above terms and conditions will be on the basis of decision taken in a meeting of the authorized officers of the Forest Department and the Forest Corporation.

7) The Forest Corporation will neither give any security for the allotted lots nor will excute any agreement separately.

8) The 'Rawanna' book printed by the Forest Corporation will be permitted to be used after their registration in the Range Office.

Labour And Organization

The U.P. Forest Corporation from the very beginning aimed at removing the prevalence of Bonded labour and at putting an end to the exploitation of labour done by the forest contractors. In this connection, the Corporation in the beginning arranged the labour one year in advance for felling, conversion and transportation. The Forest Corporation also employed the 'Dainik Vetan Bhogi' labour. In 1975-76, the Forest Corporation also employed the Field labour directly. At certain places, the complete work of exploitation was given to the Forest labour cooperatives. These cooperatives were further encouraged by the Forest Corporation, by providing the facilities of ration. The Forest Corporation also provided them (labour) ration at subsidised rates with the subsidy being Rs. 3.00 to Rs. 4.00 per labour per day.

The U.P. Forest Corporation also proposes to provide free housing facilities to the labour at the sites of work and at depots. They are also being provided with the facilities of entertainment like games, radios, etc. The Corporation also proposed to start a kitchen (instead of providing ration), with the consent of labour which could provide labour good and nutritions food. This will help in increasing the work of the Corporation on one side and on the other, it will increase the efficiency of labour. The Corporation is also making it a point to provide more than minimum wages. The Forest Corporation has also undertaken the responsibility of extraction of tender patta, which has helped in curbing bonded labour. To encourage production of tender patta, the Corporation has given flour at low prices and have distributed free Dhotis to provide additional incentive to labour. The labour is either hired on rate contract basis or on the basis of cubic meter exploitation.

The Forest Corporation has made all possible efforts to provide training to its officers and labour. The Corporation has made arrangement of providing technical training of logging to its Officers at the Forest College. The Corporation is also getting the training of class III employees i.e. Assistant Logging Supervisors/Logging Supervisors doen at the Rangers College, Haldwani. Besides this, the Officers/Officials are also required to attend short term training courses at Logging Development Institute of Government of India at Dehradun, in order to improve their logging expertise.

The Forest Corporation has the following organizational structure to undertake the complete processing of Forest produce at the site of exploitation :

DIVISIONAL LOGGING MANAGER

|

LOGGING OFFICER

|

LOGGING SUPERVISORS

|

ASSISTANT LOGGING SUPERVISORS

|

SCALERS

|

FIELD ASSISTANT

|

LABOUR (FELLING, CONVERSION, TRANSPORTATION).

The complete felling and logging operation under the control of Divisional Logging Manager, who is assisted by the Logging Officer. The Logging Officer is followed by logging supervisor and Assistant Logging Supervisors. These Logging Supervisors, supervise the complete felling and logging operation. the Supervisor is assisted by Scaler, who keeps The complete record of forestry operation and of the forest produce. Then comes the Field Assistant, who assist the Scaler and who serves as a link with the labour. And last of all, comes the labour, which in fact performs the felling operation.

Progress of Processing by U.P. Forest Corporation

The U.P. Forset Corporation started working in the year 1975. In the beginning, there was lack of staff with the Corporation, so it decided to get the exploitation done by the Forest Department staff. Due to shortage of staff, the Forest Corporation decided to work in East Almorah, West Almora, Tehri, Uttarkashi and Yamuna divisions. But could only succeed in working in the Uttarkashi division where it was already working, because labour could not be arranged in advance. In the plains, the Forest Corporation worked in some parts of Pilibhit and South Gonda Divisions. In this year, the Corporation could only produce 25,960 cu.m. of timber and 43,000 cu.m. of firewood.

In the year 1975-76, the Forest Corporation was handed over to the exploitation of South Kheri division and the extraction of Chir from Star Paper mills due to conflicts. The Forest Corporation was asked to extract the produce (Chir) to Haldwani and then hand over to the Star Paper Mill. In 1975-76, Forest Corporation extracted 81,288, cu.m. of timber and 80,794 cu.m. of firewood from the allotted area. In 1976-77, the Forest Corporation further increased the field of working to Khair extraction of Shiwalik and Western Circles. The Forest Corporation also undertook the responsibility of exploitation of timber used for toy-making and met the demand of U.P. Construction Corporation and Artificiasl Limb Manufacturing Corporation, Kanpur. In 1976-77, the outturn by the Corporation increased to 134053 cu.m. of timber and 130818 cu.m. of firewood.

In 1977-78, the Corporation was handed over the exploitation of Deodar. Besides the exploitation of Chir in Almora, the Corporation also worked in three timber lots of Almora and Khair forests of Western Circle. The outturn of timber and firewood both decreased to 130450 cu.m. and 126406 cu.m. respectively in 1977-78. But the Corporation also worked 21300 score of bamboo. In 1978-79, the processing by the Corporation further increased in the hilly as well as plains regions. In hills, the Corporation started working in Ranikhet, Bageshwar, Badrinath. Nainital and Chakrata and in plains in the division of Kheri, Dudhwa National Park and North Pilibhit divisions. In 1978-79, the outturn of timber increased to 229913 cu.m. and that of firewood decreased to 125332 cu.m., while outturn of bamboo increased to 98285 score.

In 1979-80, the Corporation completed the last years work of extraction which it could not complete due to shortage of Diesel. The Corporation further increased its working to the divisions of Pithoragarh, Yamuna and Garhwal Divisions. In 1979-80, the outturn of timber increased to 272822 cu.m., firewood to 147327 cu.m. and Bamboo to 176888 score. In the year 1980-81, the Corporation also started worksing in the Kalagarh, Lansdowne and Bharaich Division. The Forest Corporation also included the following regions/divisions in its field of operation.

(a) Khair exploitation of East Bahraich and West Bahraich.

(b) Shisham and Teak of Bahrawa Range of North Gonda.

(c) Exploitation of timber from Kalagarh, Landsdowne and Kedarnath divisions.

(d) Remaining area of Tehri and Tons divisions.

In this way, the following areas came under the working jurisdication of the Forest Corporation :

i) All the coniferous forests of Hilly Region.

ii) Bamboo of Western, Shiwalik, Garhwal and Tehri Circles.

iii) Khair of Western, Shiwalik, East Bahraich, West Bahraich and North Gonda divisions.

iv) Forests of Kalagarh, Landsdowne, South Gonda and Central circle.

In 1980-81, Forest Corporation exploited 305415 cu.m. of timber, 183916 cu.m. of fire-wood and 235576 score of bamboo.

In 1981-82, the Forest Corporation worked the following forest areas:

a) All species of Kumaon, Garhwal, Tehri, Central and Shiwalik circles.

b) All species of Kalagarh Division of Western Circles and North Gonda, South Gonda, East Bahraich, West Bahraich Divisions of Eastern circle.

c) Khair and Bamboo of Western circle.

In 1981-82, the Corporation extracted 478574 cu.m. of timber, 235374 cu.m. of firewood and 232739 score of bamboo.

In the year 1982-83, the U.P. Forest Corporation started the complete exploitation of timber and bamboo of the U.P. Forests, except social Forestry areas. The Forest Corporation also started exploiting the Minor Forest Produce like: Tendu Patta, Baib Grass, Bet, Raingal, Sal, seeds, Gum and Honey. In 1982-83, the Forest Corporation extracted 631481 cu.m. of timber, 650287 cu.m. of firewood, 700312 score of bamboo and also 389620 standard bag of timber patta. The Corporation also produced 11,348 metric ton of Baib grass. And it is since 1983, that the complete exploitation is being done by the U.P. Forest Corporation. The taking over of complete processing by the Corporation brought an end to the Contractor system of working in the U.P. Forests.

In 1983-84, the Corporation exploited 588840 cu.m. of timber which increased to 589154 cu.m. in 1984-85 and is estimated to be 658326 cu.m. in 1985-86. Firewood outturn decreased from 794787 cu.m. in 1983-84 to 488791 cu.m. in 1984-85 and is estimated to become 540659 cu.m. in 1985-86. Bamboo production decreased to 515712 score in 1983-84 and 465397 score in 1984-85 and is estimated to increase to 584030 score. The outturn of tendu patta increased in 1983-84 to 398920 S.Bag. but it decreased in 1984-85 to 343369 S.Bags and is estimated to become 380771 S. Bags in 1985-86. The Corporation produced 7,618 metric tons of baib grass in 1983-84, 9,665 metric ton in 1984-85 and 8,918 metric ton of baib grass in 1985. It is budgeted that the Forest Corporation will exploit 5,36,685 cu.m. of timber, 4,00,541 cu.m. of firewood, 8,00,000 score of bamboo, 3,72,500 S. Bags of tendu patta and 7,118 metric ton of baib grass.

The total amount of production/exploitation done by the U.P. Forest Corporation, since its inception is also shown by the table:

It transpires from the above table that the U.P. Forest Corporation has been very effective in changing the complete structure and pattern of forest exploitation. The Forest Corporation, by taking over the processing (i.e. exploitation and extraction) of forest produce has brought an end to the former pattern of working i.e. the contractors system, which was full of defects and drawbacks.

The Forest Corporation has developed very rapidly during the past 12 years. During this span of time, the Corporation has not only taken the complete responsibility of exploitation and extraction of Forses produce

PRODUCTION CHART OF U.P. FOREST CORPORATION.

Years	*Timber (Cu.m.)*	*Firewood (Cu.m.) Stacked*	*Bamboo Score*	*Tendu Patta S.Bags.*	*Baib Grass met.ton.*
1.	*2.*	*3.*	*4.*	*5.*	*6.*
1974-75	25,966	-	-	-	-
1975-76	81,000	-	-	-	-
1976-77	134,000	-	-	-	-
1977-78	130,450	126,406	21,300	-	-
1978-79	229,913	92,250	98,319	-	-
1979-80	272,822	147,327	176,888	-	-
1980-81	305,415	183,916	235,576	-	-
1981-82	478,574	222,857	232,739	-	-
1982-83	631,481	650,287	700,312	389,620	11,343
1983-84	588,840	794,787	515,712	398,920	7,613
1984-85	589,154	488,791	465,397	343,369	9,665
1985-86	658,326	540,659	584,030	380,771	8,918
1986-87	536,685	400,541	800,000	372,500	7,118

Source: Own compilation on the basis of Annual Report, U.P. Forest Corporation.

from the forests, but has also brought innovations in the method of working. The various achievement in the field of logging are as follows:

a) The exploitation of forest produce by the Corporation is being done purely on scientific lines following the silvicultural norms, which is helping in reducing the damage of forest produce.

b) With the introduction of 'Pit sawing Method' in place of axe in felling, the Forest Corporation has been able to reduce wastage and has increased utilization of produce from 50 per cent to 70% to 75%.

c) The damage to surrounding crop and forest produce has reduced, because felling is now being done by the technically trained staff and labour under the suprevision of logging supervisors.

PRODUCTION OF BAIB GRASS (1982-83 TO 1986-87)

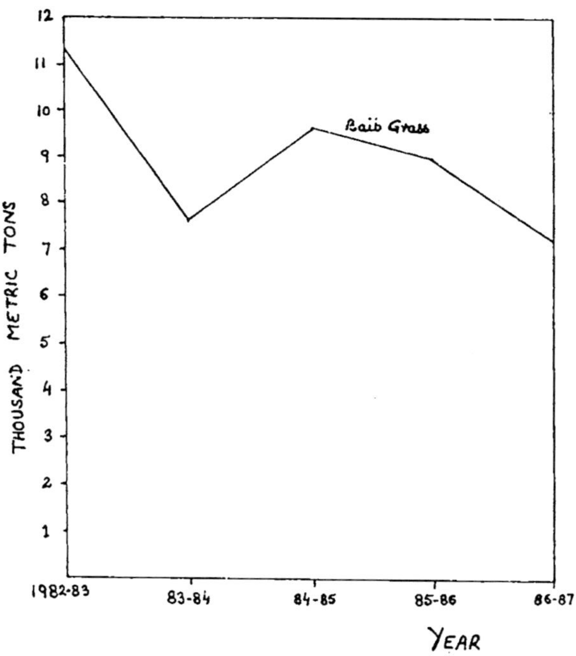

PRODUCTION GRAPH OF U.P. FOREST CORPORATION

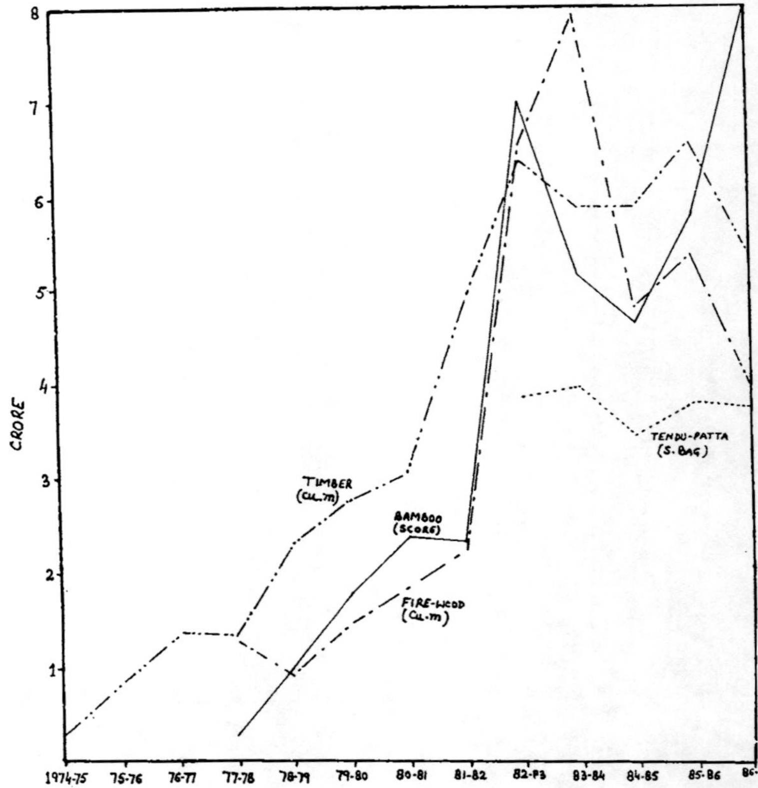

d) The forests are now being saved from illicit felling taking place during contractors system of working.

e) Damage to crop has been further reduced as the Corporation now uses better method of transportation like gravity ropeways, skyline-cranes, etc.

f) The Corporation has been able to check the exploitation of labour and has brought an end to bonded labour. The Corporation has also made efforts to improve the efficiency of labour by providing them facilities of housing, subsidised ration, etc.

Though, the Corporation has made improvement in the field of exploitation yet it has failed in bringing complete innovation. The Forest Corporation still lacks in the following areas:

i) In the field of transportation the work of loading and unloading is still being done by labour. The Corporation has not yet been able to introduce mechanised methods of loading and unloading like: Cranes, Power loaders, etc.

ii) In the field of felling and conversion, the use of mechanical saws in still needed.

iii) The Corporation has failed to establish saw mills at sites, which has affected conversion.

iv) In the lots where work is not started, the Corporation is getting allowance to do the work in next year, by paying the same royalty, which is really objectionable. For such lots, the royalty should be freshly determined and charged by the Corporation.

v) In the case of royalty calculation, the same ratio (C/CPage No 214) has been carried over the ears, which is not fully correct.

vi) Lastly, though the general price index has increased manifolds, yet the increase in royalty paid by the Corporation to the forest department has not been in accordance to the increase in price-level.

Except for these disc epancies and drawbacks, the working by

Corporation has been very beneficial to the forests of U.P

References

1. Tribhuwan Mehta, "A Handbook of Forest Utilization", 1981, p. 1.
2. "Indian Forest Utilization", Vol. I, F.R.I. Dehradun, 1970.
3. Terms and Conditions of Working by U.P. Forest Corporation.
4. Tribhuwan Mehta, "A Hand Book of Forest Utilization", 1981, pp. 29 - 31.
5. Terms and Conditions of Working by U.P. Forest Corporation.
6. Terms & Conditions of Working by U.P. Forest Corporation.

CHAPTER SEVEN

Marketing System Under U.P. Forest Corporation

Marketing of forest produce is one of the most important function performed by the forest managers. Marketing is an activity directed towards the satisfaction of human wants through exchange. It means attempting to actualize potential exchanges of forest produce with the sole-motive to satisfy human wants and needs on one hand and to earn profit to the organization on the other hand, forest products being capable of satisfying wants.

Marketing is the performance of market services by a firm. The services include exchanging of commodities, market reporting and risk taking, etc. All these market services vary with the market. Marketing would have been much simpler if all markets would have been perfect. It is the 'imperfection', which has given birth to so many marketing services and has made marketing a major branch of production. Forests also owe its liveliness to these 'imperfections'--the imperfections, being in natural bounty or in human wisdom. These imperfections give birth to cost and for the organization who is ready to bear this cost in return for revenue and profit, give rise to production.[1]

Production is the act of creating wantedness in commodities predicted power to yield satisfactions of various kinds, viz : satisfaction of form, such as lumber made from a tree; satisfaction of possession, satisfaction of place, and satisfaction of time i.e. owning the timber and having it where you want it, at the time when you want it. All these satisfactions are equally vital and it is towards them that marketing is directed.[2]

An organization secures resources from one set of market i.e. the place of origin, converts them into useful products and then trades or sells them in another set of market. It is this selling part of an organization which relates to the marketing aspect of that particular product. Thus, marketing management occurs in an organization due to any of the selling activities or activities related thereto. The task of marketing management is of regulating the level, timing and character of demand, which may help the organization in achieving its objectives. Marketing management has been defined as demand management.[3] It has also been explained as a whole business seen from the point of view of its final results, i.e. from customer's point of view.[4]

Selling has been a very old subject while marketing is relatively new. Marketing is a high order integration of man separate functions of the organization such as selling, advertising, market research, new product development, customers service, physical distribution and so on. In the business world, marketing simply means finding out an outlet for commodities. Thus, marketing involves manifold services that help selling and buying.

Marketing in the forest economics also has relevance as forests produce variety of commodities in India. No other country can boast of such a tremendous variety. Physical composition varies from a single species pure forest, to multi-species, multistoreyed forests. And these forests produce a myriad of forest products, classified under two principal groups, the timber products and the non-timber products i.e. minor forest produce. The central theme of our explanation will be based on timber and related products.

Marketing in forestry economy is more important as it has to fulfil certain social obligations which are related to conservation of the forests and maintaining the level of production of forest goods. Thus, marketing is a process of planning and managing sales in order to marry demand and supply to the best advantage of the forest entity and enterprise.[5] It is not simply the negotiating arrangement on prices and deliveries made, but it is also concerned with short-term and long-term yield regulations in the forests.

American Marketing Association has defined marketing as "the performance of business activities that direct the flow of goods and services from producer to consumer or user.[6] This definition fails to point out otherways of looking at marketing. Among the many views of marketing, the following five concepts seem to stand out :

1. Distribution of goods and services concept.

2. Delivery of standard of living concept.

3. Creation of utilities concept.

4. Generation of revenue concept.

5. System approach.

Distribution of Goods and Services Concept

This is a traditional marketing view, which has been in practice for many years. This covers all activities like transportation, grading, selling and all the efforts of whole-sellers and retailers. The use of this traditional approach creates many problems as it fails to include decisions regarding/ pricing, channels of distribution and product planning.

Delivery of Standard of Living Concept

Paul Mazur's[7] delivery of standard of living concept is a consumer oriented concept, which states that the aim of marketing is to satisfy the desire of the people in society for goods and services. Though Mazur's concept provides an excellent philosophical approach to marketing but it fails to clarify the nature of activities to be included in it. If it is taken in broader sense, the sphere of marketing operation will include everything, thereby obscuring the nature of activities.

Creation of Utilities Concept

According to traditional economist, the creation of utility viz. form, place, time and ownership, determines value of goods. The marketing activities strongly affect each of these utilities. The product planning and development creates form utility. Movement of product from production centres to the place where customer can buy creates place utility. Channels of distribution and making right goods available at right time creates time utility. Promotional activities create ownership utility by getting the goods into the hands of people who have the most use of them. Thus, marketing activity is directed to the creation of these activities in order to create value.

Generation of Revenue Concept

The generation of revenue concept is company oriented i.e. from the stand point of the firm. Marketing generates revenue at a cost which will allow the firm to earn profit from the operation. This concept proves all income generating activities as marketing action. Thus, marketing refers to creation of marketable products to exchange them in the market place with revenue.

Systems Approach

Marketing is seen as a sub-system of the business which conducts all activities relating to demand side. The society receives the goods and services through this sub-system. Marketing involves in getting the right goods to the right place at the right time for the least cost. One of the major function of marketing is concentration, dispersion, transportation and storage of the goods for selling.

Marion Clawson, an eminent Forest Economist, incharge of U.S. National Forests recommends certain actions to improve the marketing aspect of forests. To quote him-

> "Law specifies that timber shall be sold competitively, which means at bids. To avoid having a single bidder buying timber at very low prices, Forest service makes elaborate timber appraisals, which then becomes the minimum acceptable bid for stumpage. When there is active bidding the price often go far above appraised price. One major reason in addition to any errors in the process, is that the appraisal price allows full depreciation of timber processing--plus a reasonable profit margin."[8]

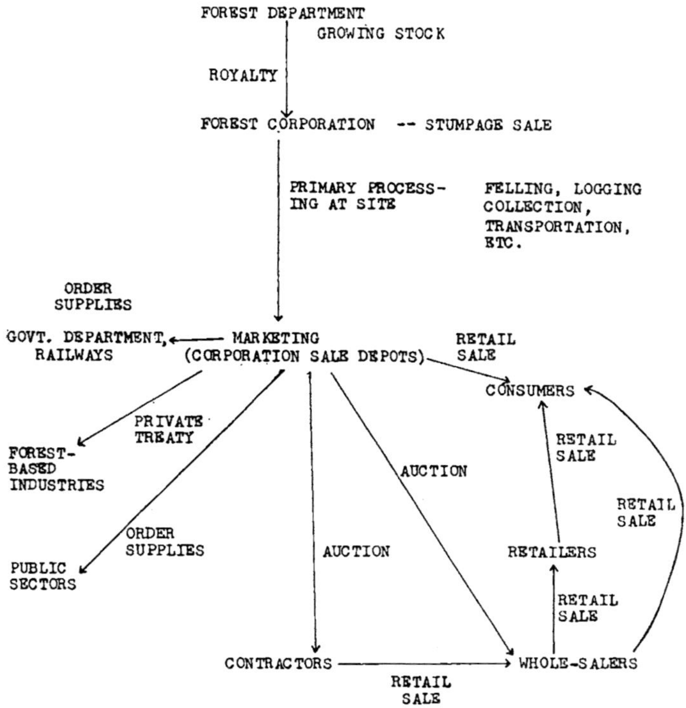
FOREST DEPARTMENT
GROWING STOCK
ROYALTY
FOREST CORPORATION -- STUMPAGE SALE
PRIMARY PROCESSING AT SITE
FELLING, LOGGING COLLECTION, TRANSPORTATION, ETC.
ORDER SUPPLIES
GOVT. DEPARTMENT, RAILWAYS
MARKETING (CORPORATION SALE DEPOTS)
RETAIL SALE
CONSUMERS
PRIVATE TREATY
FOREST-BASED INDUSTRIES
RETAIL SALE
AUCTION
RETAIL SALE
ORDER SUPPLIES
PUBLIC SECTORS
AUCTION
RETAILERS
RETAIL SALE
CONTRACTORS
WHOLE-SALERS
RETAIL SALE

"If the Forest Service, by employed labour force or by contract with loggers, were to cut trees, transport them to a reasonably central place, and sell them by auction, the whole elaborate and expensive timber appraisal process could be avoided, better net price could be obtained for logs, and the agency would have better control over the methods of logging and their impact upon the environment."[9]

Of the above two paras, first para is almost no longer applicable to Indian Forestry since recent past. Second para is what Indian Foresters are doing since the Nationalization of trade in forest products and after the formation of Forest Corporations.

The nationalization of timber trade in India completely changed the marketing system. The entire marketing of forest produce came in the hands of Forest Corporations, thus resulting in the end of contractors system. With nationalization the complete monopoly of production and marketing came in the hands of Forest Corporation. In Uttar Pradesh, U.P. Forest Corporation started marketing the forest produce.

The U.P. Forest Corporation started performing marketing of forest produce with a change and based its system more or less on the principle of Marion Clawson. The Forest Corporation shifted its marketing from stumpage sale from forest site to sale of forest produce, after primary processing i.e. after felling and conversion, from sales depots of the Corporation. The complete marketing structure which exists in the U.P. Forest Corporation is shown with the help of the following flow chart :

From the above chart it is very clear that the Forest Department has the control over the Growing Stock of the Forest. The department transfers the forest produce to the Forest Corporation. The Corporation after processing brings the produce to the Sales-depots of the Corporation from where it performs the marketing of forest produce. The complete marketing system under U.P. Forest Corporation from the point of entry of forest produce in the sale depot till its exit is discussed below in detail.

The U.P. Forest Corporation handles its complete marketing from its sales depots located at various places. The full forest produce is transported to these sales where it is stored pending its final disposal by

(1)

WATCH TOWER OF THE SALES DEPOT

sale. These sales depots are generally having large area to facilitate proper and enough storage facility. These depots are either near the market or are centrally located and have the facility of easy transportation. The following factors are kept in mind while deciding the site for sales depots :

i) The depots should be situated near the market or at central places where the demand for forest produce exists.

ii) The depots should have large area for handling timber till its final disposal.

iii) The area of the depot should be free of fungus, termites, pests and insects and should have proper drainage system, so that the stored timber may not be harmed or adversely affected.

iv) The depots should be established/located in open, where there is no danger of fire. The depots should also have their fire-fighting squad.

v) The depots should have proper security arrangements like watch towers, etc. to check theft, etc. Photo 1 shows a watch tower in one of the sales depot of Forest Corporation.

vi) There should be proper facilities of loading and unloading in the depot and also of easy transportation.

vii) Lastly, there should be proper boundary of the depot and it should also hve housing facilities for the staff of sales depot.

The U.P. Forest Corporation has opened a large number of sales-depots throughout the State to perform its marketing function. These sales depots are generally near markets and at central places and also have large capacity to handle timber e.g. Kursi Road Depot, Lucknow has the capacity to handle 20000 cu.m. of round timber. The Corporation pays the lease rent to the Forest Department, at the rates specified by it for each and every depot established by the Corporation in the Forest area.

The Forest Corporation opened a large number of depots from time to time for marketing the forest produce. The number of sales depots opened by the Corporation in various years is shown by the following table:

Year	*No. of Sales Depots*			
1974-75	6			
1975-76	12			
1976-77	22			
1977-78	23			
1978-79	21			
1979-80	23			
1980-81	28			
1981-82	35			
1982-83	72	Auction	43	Retail.
1983-84	81	Auction	53	Retail.
1984-85	102	Auction	76	Retail.
1985-86	106	Auction	98	Retail.
1986-87	109	Auction	100	Retail.

Source : Own compilation on the basis of Annual Reports of U.P. Forest Corporation.

From the above table, it is very clear that in the first year i.e. 1974-75, the Corporation only had 6 Sales Depots which increased to 12 in 1975-76 and then to 22 in 1976-77. The number of Sales depot further increased in 1977-78 and became 23 but in 1978-79, the number were reduced to 21. It is since then that the number of Sales depot constantly increased and became 23 in 1979-80, 28 in 1980-81 and 35 in 1981-82. The number of sales depots though were increasing, yet the Forest Corporation felt that it is unable to perform its marketing in the best possible manner. Thus, the Corporation decided to open more number of sales-depot so that it may be able to market more produce and may also be able to help the consumers. The Corporation from 1982-83, decided to open both Auction Sales depots and Retail Sales depots. The Corporation opened 72 auction and 43 retail sales depots in 1982-83. The number of auction depots increased to 81 in 1983-84, 102 in 1984-85 and 106 in 1985-86, while the retail depots became 53 in 1983-84,

76 in 1984-85 and 98 by 1985-86. It is estimated that the number of sales depots in 1986-87 will be 109 auction and 100 retail sales depots.

The above trend of increase in sales depots is depicted by a graph:

The above trend of increase in sales depots is depicted by a graph :

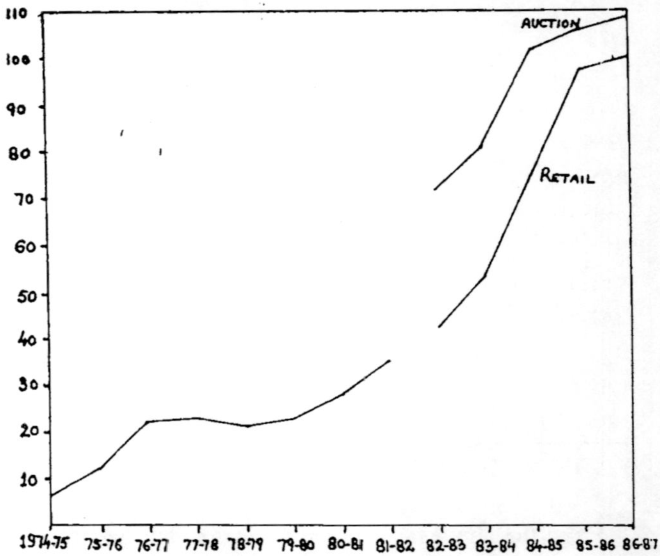

The complete list of Sales-depots : auction, retail and commission agents are given in Appendices-IV, V and VI respectively.

These sales depots are in the charge of a Depot Officer, who is of the rank of Assistant Logging Supervisor or Logging Supervisor. These Depot Officers are assisted by Scalers and Field Assistants. Besides them, there are Chowkidars and Gatemen to keep a strict vigil on the forest produce present in the depot.

The complete procedure of marketing of forest produces by the U.P. Forest Corporation, from the point of entry to exit of produce from sales depot is discussed below :

Before the forest produce enters the sale depot, it is stopped at the entry gate, where the proforma-2.1 is taken. At the entrygate, first all entries and details (like number of Truck, number of logs, grade, etc.) are completed in the import register, commonly known as 'Ayat Register'. After completing the details and checking the produce as per proforma, the truck carrying the produce is permitted inside the Corporation's Sales Depot.

After entry in the depot, the unloading of forest produce i.e. logs is done. Generally, for the purpose of unloading three to four labours are used, who with their co-ordinated efforts are able to gradually unload the produce. The process of unloading is shown in the photograph. The photograph (2) shows that the truck loaded with produce has entered the depot and the labour has climbed on it to unload it. Photograph No.3 shows the labour gradually sliding the logs to the back of the truck and photograph No.4 shows the logs being finally dropped down on the ground which completes unloading of produce.

After the unloading of forest produce is over, the counting is done and then the proforma 3.3 is issued by the depot officer. This proforma serves as a proof of the total number of timber pieces received at the depot. When the initial formalities are over the correct measurement of the produce is done and noted. If there is any difference in the stated measurement and the measurement done at depot, then before returning fthe proforma 2.1 to the same depot (i.e. the depot from where the produce is received), the required corrections are done. The complete details of the produce received is also recorded in the depots register and the Depots-Monthly proforma for receiving produce at depot, i.e. Form-3 (Jha). This form has four copies as follows: (i) Yellow for the Divisional Logging Manager, (ii) Green for the Regional Manager, (iii) Pink for sales officer, and (iv) White for office. This proforma has nearly all the columns which are necessary.

At this stage, the depot officer also gets the entry completed in the store register of the depot.

When the entire entirie/details recording is over the logs are sorted out according to species, grade, length, and girth wise basis. The

LABOUR READY FOR UNLOADING OF TRUCK

SLIDING OF LOG FROM TRUCK

(4)

FINAL DROPPING OF LOG

(5)

PLOT OF SAGON ROUND TIMBER

(6)

PLOT OF BALLI (POLES)

(7)

PLOT OF BET (CANE)

following photograph show the different forest produce sorted separately as follows:

a) Photograph - 5 : shows the plot of Sagon round timber.

b) Photograph - 6 : shows plot of balli (poles).

c) Photograph - 7 : shows plot of bet (cane).

At certain places, where saw-mills are also present in the Corporation Sales Depots, the round timber is also converted into sleepers of sawn timber of even and odd lengths, keeping in mind that minimum wastage should take place in such conversion. In the case of Sal and Asna, before converting the logs into sawn timber, debarking is done with the help of debarking spade. The conversion of logs takes place into following different forms, namely :

a) Square or balk--these are with or without waves.

b) Beams--these are rectangular in shape.

c) Sleepers for Broad gauge an meter gauge Railway lines.

d) Planks.

e) Scantling- having square or rectangular shape.

f) Lastly, firewood, which is the left over of logs.

The conversion of logs into various forms take place with the help of mechanical saw-mill. Presently, these saw-mills are owned by private individuals and are working on rate contract basis. The total number of saw-mills working under U.P. Forest Corporation from the year of inception are given in the following table:

Year	No. of Saw-Mills
1974-75	2
1975-76	13
1976-77	15
1977-78	17
1978-79	18
1979-80	17
1980-81	16
1981-82	17
1982-83	17
1983-84	16
1984-85	17
1985-86	18
1986-87	18

Source: Own compilation on the basis of Annual Reports, U.P. Forest Corporation

The table clearly shows that in the first year there were only two saw mills which increased to 13 in 1975-76 and to 15 in 1976-77. But in the year 1977-78, the number of saw-mills increased to 17 and it is since then that the number has been more or less stable. In the year 1985-86, the number of saw-mills increased to 18 and continued to be in operation in 1986-87 also. It is very important to note here that the number of saw-mills was insignificant in relation to the total outturn of forest produce by the U.P. Forest Corporation. Secondly, these saw-mills were owned by private parties and were working on contract basis.

The trend of increase in number of saw mills in clear from the graph. The number of saw-mills increased to 18 till 1978-79 and then decreased in next two years i.e. by 1980-81 became 16. In 1981-82, the number again increased to 17 and then fell to 16 in 1983-84. In the next two years the number of saw-mills increased to 18 and became constant at 18 in 1985-86 and 1986-87.

The conversion work is performed in the saw-mills. The complete ;process of conversion of round timber into sawn timber is explained with the help of photographs as follows:

The round timber is taken to the saw mill (in the case of Sal and Asna, first debarking is done with debarking spade). The photograph No. 8 shows the saw-mill from inside where the round timber is lying for conversion as is clear from photograph No.9. The labour takes this round timber to the mechanical saws for the purpose of conversion. This log is put on a long flat movable top. This top gradually moves towards the saw, thereby causing the conversion to take place, as is clear in photograph No.10. When the complete conversion is over, the sawn timber is received, which can be clearly seen in the photograph No.11, while the remainder goes as firewood. When the conversion is over, the sawn timber and fire-wood are separately stacked, as shown in the photograph No.12 and No.13 respectively.

In stacking the sawn timber, due care is taken. The place/site of stacking is first cleared of weeds, decayed wood, rubbish, etc. Then foundation is put to raise the stack. For this purpose sleepers or scantlings are used (concrete foundations would be better); while laying the foundation due care is taken keeping in mind the direction of wind. After laying the foundation, the material to be stacked is sorted out according to the length and thickness. Then the stacking is done keeping the longer length timber at the bottom and shorter at the top. Due care is taken in stacking and it is seen that timber of same thickness are stacked in the same layer. Carelessness in stacking or improper stacking will result in timber becoming warped and crooked.

The dimension of the stacks may vary only in length but in width, it should not be more than 1.5m. The stacks are arranged in such a manner

(8)

INTERIOR VIEW OF SAW MILL

(9)

TIMBER LYING FOR CONVERSION

(10)

LOG READY FOR CONVERSION

(11)

CONVERSION OF LOG INTO SAWN TIMBER

that they do not mix together. Generally, two to three feet distance is kept between stacks. The stacking of sawn timber, as done by labour is clearly depicted in photograph No.14, while thé photograph No.15 shows the sawn timber stacked separately, in separate stacks. The stacks are generally made of 8 to 10 cu.m. timber i.e. a truck load. In case of auction of timber to retail consumers, the lots are of small sizes. The timber stacked in the depot is numbered and measured and these details are also put on the timber (sawn).

In the depots of Lucknow (Kursi Road) and Bareilly, the seasoning plant is also available. In these depots, the seasoning is done. Seasoning of timber refers to removing of moisture of timber in a controlled manner. In fact, seasoning of timber starts from the time the tree is felled to ground. Seasoning of sawn timber is done with the following objectives:

a) to reduce risk of fungal and insect attack.

b) to reduce sweight.

c) to avoid seasoning defects lsike shakes, splits and cracks, etc. after the timber is put to use.

Seasoning though can take place in the open under the sun and heat waves but it is not preferred because in these cases the rate of drying is very high which causes the timber to split, crack and bend. And that is why timber is stacked at a place where the sun rays and heat waves may not directly affect the timber. Seasoning by the help of seasoning plant is done as follows:

The sawn timber is stacked on the trolley and then this trolly is taken inside the seasoning plant (Solar Kiln). When the trolley has reached in, the door of the plant is closed and fans opened. These fans help in rotating the hot air inside the solar Kiln and also in keeping the rate of drying under check. The photograph No.16 shows the stack of wood inside the seasoning plant and photograph No.17 shows the seasoning plant (Solar Kiln) in operation. When the seasoning is over, the timber is taken out and stacked.

After completing stacking, the U.P. Forest Corporation conducts the sale of forest produce. At present, the U.P. Forest Corporation has the following marketing system for the sale of its produce. This is as follows:

i) By private Treaty.

ii) By Retail Sale.

iii) Supplies to Government, public sector undertakings and industries in private sector.

iv) By acution.

1. Private Treaty

In this system, the corporation supplies timber of special size, grade and specification to the small entreprenuers and consumers as per their requirements. In this the managing committee gives the permission of sale. The sale is conducted at a price 20 per cent above the floor price or at the price collected in the last auction, which ever is greater. The small entrepreneur selects the timber. The timber is then measured and the value is determined. This value is told to the entrepreneur, who deposits the amount in the bank and gets the bank draft made. The entrepreneur then gives this bank draft to the sales officer, who finally issues the permit of export 'Nikasi', thereby entitling the pruchaser to transport the produce.

II. Retail sale

In retail sale, Forest Corporation sells its forest produce to the consumers for domestic use. In this, a fixed amount of produce is sold to the consumer. The Corporation sells bamboo, poles and sawn timber at a price 20 per cent above the floor price and fuelwood, coal, planks, etc. at the floor price to the consumers. The Corporation also supplies fuel wood to the consumers on ration card at the rate of Rs.40/- per Qtl. It also provides firewood to the employees of forest department and forest corporation at the concessional rate of Rs.20/-per Qtl. upto 4.5 quintals per year. The photograph No. 18 & 19 show retail sale plot of sawn timber at Kursi Road, Lucknow and photograph No.20 shows plot of fuel wood for retail sale. The retail sale is done on the length and cross-section basis as per the choice of cunsumer. Then the bill is prepared, the payment of which is done by the consumer through Bank Draft, payable to the Managing Director.

(12)

STACK OF SAWN TIMBER

(13)

STACK OF FUEL WOOD

(14)

STACKING OF TIMBER BY LABOUR

(15)

SAWN TIMBER SEPARATELY STACKED

(16)

STACKING OF WOOD INSIDE SEASONING PLANT

(17)

SEASONING PLANT IN OPERATION

(18)

RETAIL SALE PLOT OF SAWN TIMBER

(19)

INTERIOR VIEW OF RETAIL SALE PLOT

(20)

RETAIL SALE PLOT OF FUEL WOOD

III. Supply of timber etc. to Government Department and Corporations in Public Sector and various indutrial Units in Private Sector :

The Corporation supplies timber, firewood, etc. directly to various Government Departments and Corporations in the public sector. Though the supplies are directly made yet the Corporation does not enter into any agreement with these agencies. The total supply, which is to be met is decided by the 'APEX' Committee of the Officials of Forest Department, the Corporation, in Government and the Representatives of various sectors and industries. The Forest Corporation has so far supplied timber, etc. to the following Government Departments namely: Railways, BHEL UPSEB, U.P. Rajkiya Nirman Nigam, various sugar mills, Hindustan Copper Limited, ALIMCO, Garhwal Mandal Vikas Nigam, P.W.D., etc. The Corporation supplies to these departments, timber at a price 20 per cent above floor price, firewood, sleepers, planks, etc. at floor price. Besides this, raw material (soft wood) is also supplied to various industrial units in the private sector (e.g. Plywood, Match Wood, Pencil, Toys, Sports-Goods, Paper Mills Furniture, Packing Case, etc.) as per the allotment made by the Government and at rates fixed by the Government (i.e. subsidised rates).

IV. Auction

The Forest Corporation sells majority of its forest produce by Auction Sale. It sells timber in both forms i.e. round timber (logs) and sawn timber and other forest produce on the specific dates, predecided for each sale-depot which all prospective buyers can have from the office and are also availabie at sale-depot. A list of these dates, species and names of sale-depot is also hung at the sale depot's office as shown in photograph No.21.

The Forest Corporation prepares lots of suitable sizes (i.e. approx. a truck load or 8 to 10 cu.m.) and then auctions them. The following photographs show lots of different species like :

i) Photograph No. 22 - shows plot of round timber.

ii) No. 6 - shows Balliplot.

iii) " No. 7 - shows Bet (cane) plot.

iv) " No. 15 - showplot of sawn timber.

The Forest Corporation, on the predecided date finally auctions the produce. The Corporation follows the following procedure or conditions of sale of timber, etc. by auction :

UTTAR PRADESH VAN NIGAM - CONDITIONS FOR AUCTION SALE : ORDINARY CONDITIONS

1. The auction of forest produce will be done in lots of a truck load on lump-sum contract system. Every purchaser will have to deposit Rs. 1000/- as gate money before attending the auction. Only the gate money depositor contractors can bid in their names. No relaxation will be allowed in this respect.

2. Although, measurement of every piece included in the lot is correctly depicted in the sale list as far as possible, yet the purchaser is advised to see the lot carefully before bidding. If any purchaser doubts the measurements of pieces (except fuel wood) entered in the sale list and desires re-measurement, then that purchaser will give a written application to the auctioner and deposit alongwith it Rs. 50/- or one per cent of the price of the lot, whichever is more. If difference of over 2 per cent is found between the actual remeasurements and previously recorded measurements, then the sale of the lot will be cancelled and all the amounts deposited by the purchaser in respect of that lot will be returned to him. But if the aforesaid difference be only 2 per cent or less, then the sale will be considered valid and the security will not be returned.

3. After the sale of the lot, no complaint about the grade of timber will be entertained from the purchaser. The purchasers should bid only after seeing the grading.

4. The auction of lots will ordinarily be on the basis of highest bid. But the Corporation will not be bound to accept the highest or any other bid. The Corporation will have the right to reject the highest

or any bid without assigning reason. Immediately after the auction of the lot the purchaser will have to deposit 10 per cent of the total price of the lot as security. Security amount will be deposited either cash or through a bank draft on any nationalised bank in favour of the Managing Director, U.P. Forest Corporation. The Bank collection charges payable on that will not be demanded from the purchaser. This expenditure will be borne by the Forest Corporation. In the event of failure to deposit security the sale of the lot will be cancelled and the gate money will be confiscated.

5. On approval of sale of the lot, a bill in account form No.3-B-1 will be issued in purchaser's name which will be sent to him through registered post. The amount payable as shown in the bill shall have to be deposited within 24 days (but before lifting the material) of the issue date of the registered letter (leaving the day of despatch). If the bill has been delivered to the purchaser by hand, then the amount payable shall hve to be deposited within 21 days (but before lifting the material) of the receipt of the bill (leaving the day of receipt). This payment shall have to be made in the depot office through bank draft drawn on any nationalised bank in favour of the Managing Director, U.P. Forest Corporation. If the bank draft is received even upto 3 days after the date mentioned in the bank draft, then it will be deemed to have been received on the date written on the bank draft. The payment of due amount of sale price will not be accepted in cash or through pay-in-slip of the bank.

6. Sales Tax on the sale price will not be charged separately from any purchaser. The sales tax due will be deemed to have been included in the sale price.

7. The maximum limit for the approval of sale by auction of all kinds of lots will be 40 days from the date of auction. If the period of 40 days from the date of auction expires and the purchaser does not receive in due time the information of approval of sale (through Account Form No. 3-B-1), then the purchaser can refuse to buy (accept) the lot and in such a case the security deposited by him for the lot will be refunded to him on his demanding it.

8. If the purchaser makes full payment within 10 days (15 days for conifer species in hilly areas) of despatch of the bill or within 7 days (10 days for conifer species in hilly areas) if the bill has been received by hand, then a rebate of 1% (one percent) in the sale price of the lot will be allowed to him in plain region. In hill region for conifer species this rebate will be 3% (three percent).

9. If the purchaser does not make payment of the remaining amount within 24 days (excluding day of despatch) of the bill or within 21 days (excluding the day of despatch) of the receipt of bill by hand then a late fees will be charged upto next 7 days at the daily rate of 0.25 per cent on the amount due. If the contractor does not make the payment of due amount even after the permissible period of 7 days for late fees, then his security will be confiscated and the auction of the lot will be cancelled.

10. Partial or full 'Nikasi' of the wood/timber of the lot will be allowed to the purchaser only after he makes payment of the full price.

11. Regarding payments, only the receipts duly signed and stamped by the Depot Officer and the Depot Accountant will be considered valid.

12. It shall be compulsory for the purchaser to do 'Nakasi' of the lot from the depot within 45 days from the date of despatch of the bill (excluding the date of despatch) or within 42 days of the receipt of the bill by hand. If due to any reason, the purchaser cannot do full 'Nikasi' of the material from the depot within the aforesaid period, then till 15 days after this period 'Nikasi' will be given with late fees. The late fee will be for each day of delay, 0.25 per cent of the cost of remaining material of the lot lying in the depot. If the purchaser fails to do 'Nikasi' of the material of the lot from the depot even before the expiry of the late fee period (of 15 days), then;

   a) the Forest Corporation will confiscate all the amount deposited by the purchaser for that lot, and

   b) the Forest Corporation will become the owner of all the material of the lot kept in the depot, and will be free to disposed it off either by auction or through other means and the purchaser will not be entitled to any kind of compensation, etc. for this.

13. The purchaser will be allowed 'Nikasi' from the depot only between the sunrise and the sunset.

Special condtions

14. *Timber*--Lots of all kinds of timber as well as khair will be made of estimated one truck-load and the measurement of every piece of the lot will be depicted on that piece. Lots of all kinds of timber will be sold by volume and not by weight.

15. *Fuel wood and 'Pharras'*-- Auction of fuel wood and of 'Pharras' (off cuts) coming from swaing will be by 'chatt' volume after making about one truck load of 'chatta' (7 m x 3 1 m). Every 'chatta' will constitute a lot, whose bid will be taken in one lump sum.The purchaser is advised to see the 'chatta' very carefully before bidding because the Forest Corporation does not accept any responsibility for any difference creeping into the measurement of 'chatta'. There is no provision for remeasurement in the case of firewood lots.

16. *Khair Chunni, roots of all species, Asna bark & sawdust*--The lots of these will be made by estimated weights. Their bid will be taken per quintal only and their security too will have to be deposited according to their estimated weight.

17. The proforma bill of the sold lot will be issued only on the basis of weight depicted in the lot, and for the payment of this bill all the ordinary conditions will apply.

18. For the 'Nikasi' also from the lot, ordinary conditions will apply. In cases of 'auction of lots by weight', the loaded truck will be weighed in a weighing balance, recognized by the Forest Corporation before the authorized officials of the Corporation and on the weighment paper will sign both the authorized official of the Corporation and the purchaser. The final payment bill will be prepared only accordisng to the weighment paper so obtained, and if the due payment shown in this bill is in excess of the proforma bill then the purchaser will have to make payment of the additional

amount immediately, and only then the purchaser will be allowed to remove his truck from 'Dharam Kanta'. If already over payment has been made then the additional (excess) amount will be refunded to the purchaser according to rules.

19. The empty truck brought for 'Nikasi' will have to be got weighed by the purchaser at the 'Dharam Kanta' recognized by the Corporation in the presence of the authorized officials of the Corporation. If the weighment of empty truck cannot be done for want of a 'Dharam Kanta' nearby, then the weight of empty truck, as recorded in the registration book of the truck, will be taken for granted. In such cases when the material is less than a truck load, its weighment will be done by the beam balance avialable in the depot.

20. In case of any dispute relating the implementation of these terms and conditions, the concerned Regional Manager, U.P. Forest Corporation will act as an arbitrator whose decision will be final and binding on both the parties.

After completion of auction of forest produce and receipt of its payment by the Forest Corporation, transit permit is given to the purchaser. The purchaser is given permission to bring his truck inside the sale-depot for the purpose of loading. The loading of the truck is done by labour (i.e. concerened employee) in the presence of the sales officer or his representative. The photographs : 23 & 24 show loading of firewood and timber at Kursi Road Depot. When the loading of the truck with the produce is over, a challan No.4.5 is issued after checking the produce, in the name of purchaser alongwith the truck number and the complete description of the material. The truck carrying the produce is then permitted to move from plot to the exit check post.

The truck loaded with forest produce is checked at the gate of the sale-depot. Here the produce is again checked by the gate-keeper as per the proforma and hammered. The gate-keeper completes all the entry of the Export Register, commonly known as 'Nikasi Register', and issues a final export proforma 3.4. The purchaser is also issued a 'Nikasi' permit so that he may not face any hassles on the way i.e. in transportation. Thus, the purchaser finally takes the produce to his depot or to the market for retail-sale purpose. The photograph No. 25 shows the checking of forest produce at the gate.

(21)

BOARD SHOWS LIST OF SALE DEPOTS, SPECIES AND DATE OF AUCTION

(22)

PLOT OF ROUND TIMBER

After final export of timber i.e. sale of timber and its removal from depot is over, the Form No.8--sale, is completed in the register and the final report is sent to the Chief Marketing Manager by Divisional Sales Managers of the respective depots.

Organization/Labour

The organization of the U.P. Forest Corporation to perform the specialized function of marketing is highly professional and has developed with the passage of time. The Forest Corporation employs people having technically sound knowledge in the field of marketing, so that it may perform the sales in the most efficient manner and collect the maximum sales proceeds. To deal with the marketing of forest produce, the U.P. Forest Corporation has the following organization.

On the top, at the headquarters level is the Chief Marketing Manager, Marketing Manager and the Sales Oficer. While at the Divisional level, the Divisional Sales Manager is on the top, who is followed by Sales Officer. The Sales Officer is assisted by Logging Supervisors and Assistant Logging Supervisors. These Assistant Logging Supervisors are assisted by Junior Logging Assistants, Scalers and Field Assistants, who in turn are assisted by labourers.

The above organization can be shown as follows:

DIVISIONAL SALES MANAGER
|
SALES OFFICER
|
LOGGING SUPERVISORS
|
ASSISTANT LOGGING SUPERVISORS
|
JUNIOR LOGGING ASSISTANTS
|
SCALER
|
FIELD ASSISTANT
|
LABOURER

(23)

LOADING OF FUEL WOOD

(24)

LOADING COMPLETED

In the above organization, all the posts are filled on regular employment basis, except that of labourers. The labourers are employed on rate-contract basis or are hired on cubic-meter basis. Though the labourer is not permanently employed, yet due care is taken by the Corporation while employing labour and it is observed that the same labour (who has been in employment previously) is again employed, when required.

Progress of Sale

In the year 1974-75, the Forest Corporation extracted twisted chir from Almorah and Nainital and supplied it to Star Paper Mill, Saharanpur from the Haldwani Depot. The Corporation delivered almost 1,50,000 quintals of twisted chir timber to Star Paper Mill. The Corporation also entered into agreement with the Bharat Heavy Electricals, Hardwar to supply Chir timber worth Rs.25 lacs annually. The Hindustan Copper Ltd., Khetri (Rajasthan) also entered an agreement for supply of sawn timber and Sal poles worth Rs. 6 lacs annually. The Corporation is also supplying Railway sleepers to the extent allotted through the Forest Department.

In 1977-78, the Corporation supplied 20 per cent of the turnover directly to various public sector undertakings like Railways, B.H.E.L., Hindustan Copper Limited, Khetri; U.P. State Electricity Board, U.P. Avas Evam Vikas Parishad, U.P. Rajkiya Nirman Nigam, ALIMC., etc. and the local consumers. In addition to this, Forest Corporation is also supplying firewood to U.P. Sugar Corporation. For the convenience of public, the Corporation has opened a large number of sale-depots at various places to meet their requirements of timber and firewood. The Corporation is also supplying good quality timber at subsidised rate to the U.P. Artificial Limb Corporation, so that good quality artificial limb production can be possible.

In the year 1978-79, the Corporation supplied 24 per cent of the turnover to the public sector and the local consumers. The Corporation also supplied good quality of soft wood at subsidised rate to the U.P. Artificial Limb Corporation for the production of good quality artificial limbs. In 1979-80, the Corporation supplied timber to the various public sector units (including P.W.D.). This year, the Corporation supplied produce amounting to 22 per cent of total sale value from Sal region and produce worth 32 percent of total sale value from coniferous forest. To benefit the consumers, the Corporation has opened a number of retail sale depots from where timber of all sizes and firewood is being made available. The Corporation supplied forest produce (Sal and other hard wood) worth 1.6 per cent of total sale value and coniferous worth 2.4 per cent of total sale value, to the consumers from its retail depots.[10]

In 1980-81, Corporation supplied timber worth 12 per cent of the total sale value from sal region and timber worth 13 per cent of the total sale value from coniferous forest region to the public sector-under-takings. The Corporation also supplied various types of timber to the consumers on reasonable rates. The Corporation supplied timber worth 2.3 per cent of total sale value to the consumers.[11] In 1981-82, the Forest Corporation continued to supply timber to the public sector as before. In this year, the Corporation supplied timber for the consumers worth 2.5 per cent of the total sale value.[12] The Forest Corporation, besides auction, also sold timber, firewood, Khair, etc. to consumers through private treaty, amounting to Rupees One Crore.

In 1982-83, U.P. Forest Corporation, generally sold major proportion of forest produce through auction and the rest of directly to the consumers through sale-depots. But as the Corporation failed to get good prices of its produce by auction sale so it decided to sell some produce directly to the consumers through private treaty. In this manner, the Corporation sold forest produce (like fire-wood, khair wood, soft timber, etc.) to the consumer worth Rs. 1.25 crore. The Corporation also supplied soft timber to the 112 industrial units (match, plywood, sports goods, packing case, pencil, etc.) as per the allotment of the 'Apex Committee' at fixed rates. In this year, Corporation supplied 15,000 cu.m. of timber to these industrial units. In this year, Forest Corporation supplied timber worth 4.6 per cent of total sale value to the public sector and timber worth 2.1 per cent of total sale value to the retail consumers at reasonable rates [13]

In 1983-84, the Forest Corporation supplied timber and firewood worth Rupees two lacs to the consumers through private treaty and a sizeable amount of forest produce through retail sales. Altogether, Corporation sold timber and firewood worth Rs. 10.13 crores and Tendu patta worth Rs. 13.30 lacs. Besides this, the Corporation supplied 8,000 cu.m. of soft wood to 112 industrial units, at the fixed rates decided by the 'Apex Committee'. Thus, during 1983-84, Forest Corporation supplied output worth 6.5 per cent of the total sale value to the public sector and timber worth 1.7 per cent (approx.) of the total sale value to the consumers on reasonable rates.[14]

During the year 1984-85, the Corporation was able to market more forest produce. The total sale of the Corporation during 1984-85 is given in the following table.[15]

Method of Sale	*Collected sale value of Forest produce.*	*% to total sale price*
Auction	Rs. 6580.3 lacs	70.4
Order Supplier	Rs. 939.43 "	10.1
Retail Sale	Rs. 196.10 "	2.1
Private Treaty	Rs. 1630.87 "	17.4
Total =	Rs. 9346.43 Lakh	100.0

The U.P. Forest Corporation sold produce worth 70.4 per cent of total sale price, amounting to Rs. 6580.30 lacs through auction, produce worth 17.4 per cent of total sale value i.e. worth Rs. 1630.87 lacs through private treaty, produce worth 10.1 per cent of total sale price i.e. worth Rs. 939.43 lakh through order supplies to the Government Departments and produce worth 2.1 per cent of total sale value i.e. worth Rs. 196.10 lacs through retail trade. Thus, the Forest Corporation sold produce worth 70.4 per cent of total sale value through auction and remaining produce was sold directly. During this year, Forest Corporation supplied forest produce to various industrial units and Government Departments in the under-mentioned manner :

i) The Corporation supplied 9,000 cu.m. of soft wood to 97 industrial units (like match industry, sports-goods industry, packing, plywood, etc.

ii) The Corporation supplied 18,305 Sal Sleepers and 3,000 Chir Sleeper to Broad gauge, 40,388 Sal and 3,000 Chirsleepers to Meter gauge, and 897 cu.m. of Sal timber to the Railways.

iii) The Corporation supplied 53,000 cu.m. of firewood to Sugar Mills.

iv) The Corporation supplied 1,61,400 cu.m. of Eucalyptus pulp to the Century Paper and Pulp Limited.

v) Corporation supplied 6,500 Eucalyptus poles to the U.P. State Electricity Board.

In the year 1985-86, the overall sale of the U.P. Forest Corporation increased. The Corporation supplied directly to various industries and Government Departments in the undermentioned manner[16]:

i) 13052 cu.m. of soft wood was supplied to the soft-wood based industries of U.P.

ii) Corporation supplied Sal Sleepers to the Railways as follows :

(a) 27,261 - Broad Gauge.

(b) 56,785 - Meter Gauge, and

(c) 15,959 - Special Size.

iii) 1,08,000 cu.m. of firewood was supplied by the Forest Corporation to the U.P. State Sugar Corporation and to the sugar units of Sugar Mills Cooperative Union.

iv) 37,011 cu.m. of Eucalyptus pulp was supplied to Star Paper Mills and 97,967 cu.m. of Eucalyptus pulp to Century Paper and Pulp Mill Limited, Lalkua.

v) Corporation also supplied 18,000 cu.m. of Khair timber to 8 Katha producing units, as per the orders of the Government.

The Forest Corporation sold forest produce worth Rs. 7815.48 lacs by auction, Rs. 1826.07 lacs through order supplies, Rs. 610.34 lacs through private treaty and produce worth Rs. 256.68 lacs through retail sale.

It is estimated that the Forest Corporation will be meeting the following supplies during 1986-87 :

i) 13,228 cu.m. of soft-wood will be supplied to 81 industrial units.

ii) Corporation will supply 24,000 Broad Gauge, 50,000 Meter Gauge sleepers and 1,001 cu.m. of Sal timber to the Railways.

iii) The Corporation is expected to provide 1,00,850 cu.m. of firewood to the sugar mills.

iv) The Corporation is expected to supply 7,000 cu.m. of Khair to the Katha producing units.

The overall trend of sale of the U.P. Forest Corporation during 1974-75 to 1986-87 can be studied with the help of the following table:

Year	*Sale Price obtained in Lakh Rupees*
1974-75	63.96
1975-76	228.38
1976-77	555.72
1977-78	697.53
1978-79	823.00
1979-80	1420.00
1980-81	2342.92
1981-82	3425.62
1982-83	5924.59
1983-84	7624.10
1984-85	9164.76
1985-86	10508.57 (Estimated)

Source: Own Compilation based on Annual Report, U.P. Forest Corporation.

From the above table it is clear that in the first year i.e. 1974-75, the Forest Corporation was only able to obtain sale price of Rs. 63.96 lacs, which increased to Rs. 228.38 lacs in 1975-76 and Rs. 555.72 lacs in 1976-77. In 1977-78, the sale price obtained showed a small increase but it is since then that the sale-value of produce has been increasing at a higher rate. The sale-price obtained increased to Rs. 823.00 lakh in 1978-79, Rs. 1420.00 lakh in 1979-80 and reached to Rs. 3425.62 lakhs in 1980-81. From 1981-82, the Corporation was assigned the complete responsibility of exploitation and sale of forest produce of the whole state, so the Corporation was able to sell greater amount of produce and was also able to earn large sums of money as sale-price. The sale price obtained during 1981-82 was Rs.3425.62 which increased to Rs. 5924.59 lakh in 1982-83. The sale price further increased to Rs. 7624.10 lacs in 1983-84 and Rs. 9164.76 lacs in 1984-85. It is estimated that the sale-price obtained during 1985-86 will be somewhere around Rs.1,05,08,57 lacs. the increase in the sale value was partially because of increase in the quantity of produce sold and partially due to increase in price level.

The trend of sale-price obtained from 1974-75 onwards is depicted with the help of curve

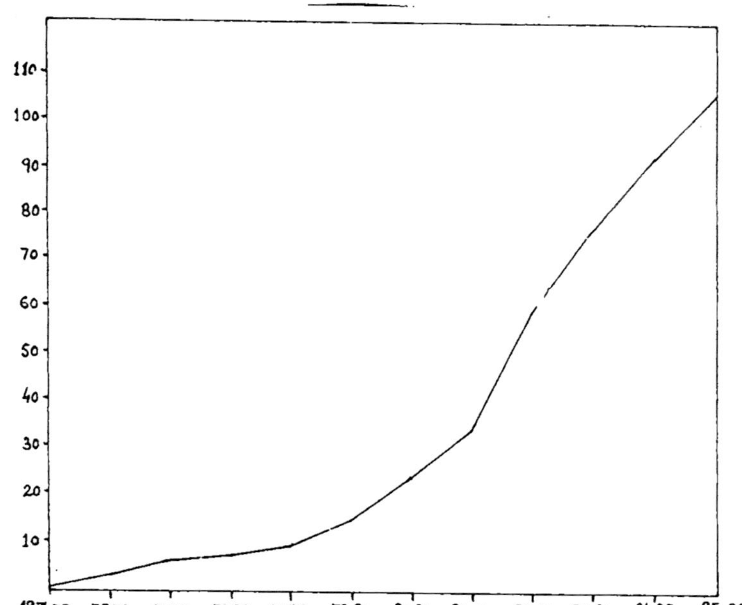

The curve shows that the sale price obtained by the Corporation has constantly increased but the increase was comparatively much slower till 1981-82. But it is since 1981-82 that the sale price obtained has shown faster increase.

By marketing the forest produce the Corporation also earned profits as given below :

PROFIT-1974--75 TO 1985-86

Year	*Profit (Lakh Rupees)*
1974-75	8.82
1975-76	48.26
1976-77	8.92
1977-78	36.25
1978-79	4.64
1979-80	29.21
1980-81	64.68
1981-82	80.11
1982-83	144.06
1983-84	190.77
1984-85	222.37
1985-86	270.41

Source: Own compilation on the basis of Annual Report, U.P. Forest Corporation.

The profit earned by the Forest Corporation kept fluctuating till 1979-80 i.e. the Corporation earned Rs. 8.82 lakh of profit in 1974-75 which increased to Rs. 48.26 lakh in 1975-76, but again fell to Rs. 8.92 lakh in 1976-77. In 1977-78 the profit of Corporation increased to Rs. 39.25 lakh but again fell to Rs. 4.64 lakh. In 1979-80, the Corporation earned a profit of Rs. 29.21 lakh and it is since then that the Corporation has been earning more and more profits. In 1980-81, the Corporation earned Rs. 64.68 lakh as profit, which increased to Rs. 80.11 lakh in 1981-82, Rs. 144.06 lakh in 1982-83 Rs. 190.77 lakh in 1983-84 and Rs. 222.37 lakh in 1984-85. It is established that the profits of Corporation will become equal to Rs. 270.41 lakh during 1985-86. The above trend of fluctuation of profit is shown with the help of the following graph:

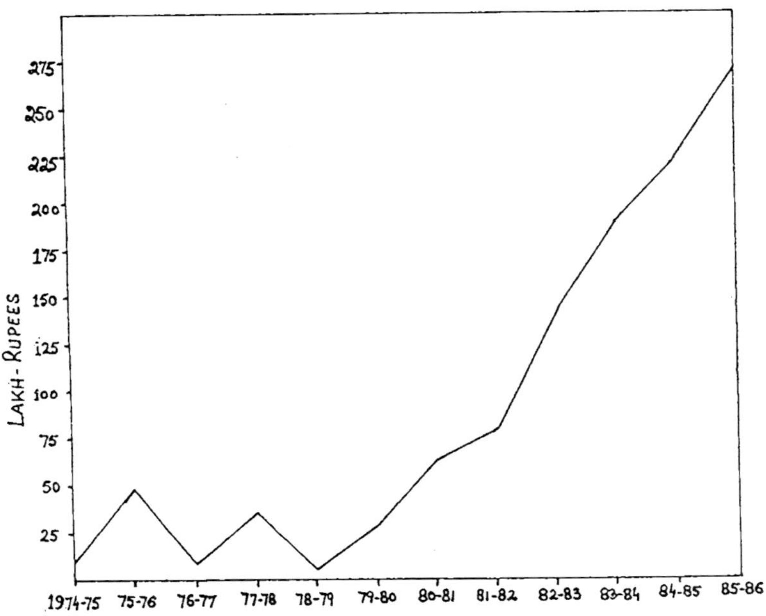

The marketing of forest produce by the U.P. Forest Corporation has envisaged the following improvements:

i) Marketing by Forest Corporation has brought an end to the conventional age old practice of marketing of forest produce by contractors of the Forest Department. This has helped in partially eliminating the middle-men in the forest trade.

ii) Forest Corporation has helped in checking pilferage, which used to take a heavy toll of timber and firewood in the former pattern.

iii) Corporation has helped in checking the concentration of economic power, which was taking place among the forest contractors.

iv) Marketing of forest produce by the Forest Corporation has helped the consumers in the following manner:

a) Corporation is supplying firewood to the consumers on subsidised rate on ration card.

b) Corporation is providing good timber on retail basis directly to the consumers at reasonable rates from its sale-depots.

v) Corporation is supplying timber regularly to the Railways, housing boards, defence and other public sectors, by department to department basis, on fixed schedule of rates.

vi) Corporation is also helping the soft wood based industrial units and paper mills by regularly meeting their demands for raw material.

vii) Corporation has played an important role in increasing the value of forest produce by selling the produce after conversion. This has helped oin maximising the return, reducing wastage, increasing the earning of the state exchequer and has also helped in making the sawn--timber of all sizes available for desired purposes.

viii) The problem of collecting revenue from the contractors has also been solved.

ix) Lastly, the marketing by Forest Corporation has also helped in increasing the employment opportunities at all levels.

(25)

CHECK-POST OF THE DEPOT

(26)

DELAPIDATED STATE OF AGRA DEPOT

Though, marketing by forest produce by Forest Corporation has been beneficial yet it is not free from certain defects as follows:

i) The sale-depots, from where the timber is being sold, are not accessible to the people. They are generally located far from cities.

ii) Certain depots of the Corporation are very poorly maintained as is clearly evident from the photograph No.26. The photo shows the Corporation's sale depot at Agra in a bad state, having no definite boundary and with the depot's office in a dilapidated state. The barbed wire used for demarcating the boundary of the Corporation can be seen lying on the ground in photograph No.27. This state of barbed wire is no constraint in the movement of trespassers.

iii) The sales-depots though may be having large area but are not having proper storing facilities. The timber is kept in the open and has to bear the heat of the sun in summer and the rainfall in the rainy season, with the result that splitting and cracking of wood takes place. Besides this, the area of the depots are not properly treated for the protection of timber.

iv) The Corporation, though a specialised institution, has still not been able to make mechanical arrangement of loading and unloading. Due to lack of facilities of loading and unloading the damage to timber takes place.

v) The corporation has failed to have its own saw-mills.

vi) The conversion of timber is done in the traditional way by illiterate labour, with no concept of mensuration and maximum yield. The Corporation should have a technical person who could make sure that the maximum yield is taking place in sawing.

vii) It has also been observed in certain sales-depots of the Corporation that the new incoming timber is marketed while the previously stacked timber continues to rot in the depot as is clearly visible in photograph No.28 & 29.

viii) The Corporation has failed to create an awareness among the

(27)

BARBED WIRE USED FOR DEMARCATING THE DEPOT BOUNDARY IN BROKEN STATE

(28)

VIEW OF AGRA SALES DEPOT

(29)

STACKED TIMBER IN ROTTEN STATE

masses. The people are still unaware of the services rendered by the Corporation. Besides this, it has also been noticed that in certain depots only a particular type of timber is available which is not required by the people; thus, proving the inability of the Corporation to meet the retail demand of the consumers.

ix) The sales-depots are not having weigh-machines, and at places where they are available, they are of traditional type.

x) There also exists a problem of making payment, because the person first has to know the amount and then go to the city to get the bank-draft made and then only he is able to get the produce.

xi) The labour employed by the Corporation is untrained and temporary, though the same labour is being employed again when required. This proves the inability of the Corporation to create permanent employment and bring about labour welfare.

xii) There are illegal trends developing at various places. The purchasers are constantly complaining that their lot number has been changed or their produce is being stolen or their produce is being changed. It has been noticed that certain strong purchasers are able to get excess loading or wrong loading done either by force or by bribing the lower ranking supervisory staff.

xiii) Lastly, there are organizational defects in the marketing system of Forest Corporation. The sales-depots are in the charge of Sales Officer, who are of junior rank. Then the sales-depots are not having proper security arrangements against fire, theft, cheating, etc. It has been observed that when major discrepancies arise in the stock of the depot, generally fire takes place there, resulting in major loss of forest produce. There should be a procedure of recovery of all losses, then only the returns of Corporation can increase substantially.

Though, the Forest Corporation marketing system is having certain defects, yet it would not be wrong to say that the Corporation has helped in eliminating contractors to a great extent and has also helped in the development of marketing of forest produce on modern lines. If the defects of the Corporation are removed, the Corporation's return wil increase and it will also be able to perform marketing in a better way

References

1. W.A. Duerr, "Fundamentals of Forestry Economics", 1960, pp. 323 - 324.
2. W.A. Duerr, "Fundamentals of Forestry Economics", McGraw Hill, London, 1960, p. 324.
3. Philip Kotler, Marketing Management : Analysis, Planning & Control, Prentice Hall of India Pvt.Ltd., 1980.
4. P. Drucker, "Management, Task, Responsibilities, Practices; Allied Publishers Pvt.Ltd., 1978.
5. D.R. Johnston, A.J. Grayson & R.T. Bradly, "Forest Planning; Faber and Faber Ltd., 1967.
6. S.A. Ralph, "Report of the Definitions Committee," Journal of Marketing, 1948, , pp. 202 - 217.
7. Paul Mazur, "The Standards We Raise": Harper & Row, New York, 1953, pp. 18 - 28.
8. A.R. Maslekar, Marketing of Forest Products in India", Indian Institute of Forest Management, 1979, p. 13.
9. Ibid, p. 13.
10. U.P. Forest Corporation, Annual Report, 1979-80, p. 11.
11. U.P. Forest Corporation, Annual Report, 1980-81, p. 9.
12. U.P. Forest Corporation, Annual Report, 1981-82, p. 15.
13. U.P. Forest Corporation, Annual Report, 1983-84, pp. 7-10.
14. U.P. Forest Corporation, Annual Report, 1983-84, pp. 7 - 10.
15. U.P. Forest Corporation, Annual Report, 1984-85, pp. 7 - 10.
16. U.P. Forest Corporation, Annual Report, 1985-86, pp. 7 - 10.

CHAPTER EIGHT

Retrospect and Prospect --A Critical Evaluation

The U.P. Forest Corporation shered in the field of Forest Management with certain laudable aims and objectives, which were expected to bring about a marked change in the ethos and concept of Forest Management. How far it has succeeded in this endeavour will be discussed alongwith marketing and related problems.

The main aims and objectives of the U.P. Forest Corporation besides the State Trading in Forest products, included the initiation of Forestry Research, the development of forests entrusted to it by the forest department, the establishment of forest based industries, the development of labour cooperative societies, the training of personnel, the engagement of educated unemployed and a host of other items.

So far as the State Trading in Forestry is concerned, the U.P. Forest Corporation has made an all out effect. Starting from the far-flung inaccessible areas it has now covered the entire forests of U.P. except the Social Forestry Divisions, (Which too are now proposed to be covered in the next 10 years). In addition to the major forest produce of timber and fuel, it has also taken up minor forest produce like Tendor leaves, Bamboos and Bhabhar grass, etc.

The establishment of Forest Corporation has resulted in the elimination of Contractors System from main Forestry operations like fellings, conversion, transportation, etc. Naturally, the evils trailing with the contractor system have also come to an end, though some people still have doubts in this aspect of Corporation's performance. The employment of contractors agency in the extraction of forest produce was considered to be the main cause of wanton destruction and over exploitation of forests. 'Chipko' movement highlighted this and accelerated the establishment of forest corporation in U.P. Apparently, the Forest Corporation seem to have achieved this objective of eliminating contractors' agency from exploitation but in today's situation the middleman become necessary components of timber and other forest produce trade, so the same contractors and middleman who were to be removed from the scene are very much there but in a new role as purchasers and agents under the state monopoly of forest produce trade. Similarly, the same labour with its old ideas and implements, are also present. There has only been a change of masters, from the erstwhile 'Maldars' (Forest Contractors) to the Forest Corporation. There has been little progress in the formation of labour Cooperatives, with the result that the damage of forest crop and wastage of timber due to use of primitive methods at the hands of untrained labour still continues. However, the labour is now in a happy position under the Forest Corporation as it is assured of minimum wages with regular payments, fixed working hours, provision of ration and certain other basic amenities which contractors always failed to provide. Still, much requires to be done towards labour welfare by way of Group-Insurance, Old-age Pensions, etc. The formation of labour cooperative societies has not kept pace with the development of the Corporation and the Corporation has even failed to utilize the existing cooperative societies.

The much needed and avowed research programmes relating to forest and forest products have not yet been undertaken by the Forest Corporation. Inadequate attention has been paid so far to R & D by the Forest Department as well. It is imperative that, to support extensive developments in the forestry sector there must be a strong backing of an Research & Development effort. Among other things, this could be directed towards developments of seed orchards and seed production areas, provenance trials, study of use of fertilizers and irrigation methods in man made forests, aerial spraying and other pest control measures, genetical studies including tissue culture, detailed soil surveys, etc.

It has been pointed out that the Corporation Officials (or for that matter even the Forest Department Officials) are not able to prepare the estimates of outturn of timber from trees as accurately as it should be. Therefore, there should be comprehensive studies 'to determine outturn factors which could help in preparing correct estimates. This is very necessary, for if the estimates are on lower side, the logging staff is apt to leave some part of the tree unconverted after the estimated outturn has been obtained, or this surplus is likely to be misappropriated.

The Corporation should welcome any outlays on R & D because there is an Income Tax rebate on this item. Another way of increasing the R & D effort would be to encourage sponsored research through universities. Yet another function of the Corporation about which so far nothing has been done is the "preparation of projects relating to Forestry within the state". Just possible, the Forest Corporation may not have been called upon to prepare such a project. Any way, this has created a wide gap between the conception and performance of the Forest Corporation. If this function be deemed integrated with the research and development programme, then in this too nothing has been done so far as would be seen in the preceding paras.

The main basic function in which no headway seems to have been made is the management, maintenance and development of such forests as are transferred or entrusted to the Corporation by the Forest Department. What to say of this, not even the reforestation of the felled areas has been undertaken by the Corporation, although this is one of the avowed objective of the Corporation. Unless the Forest Corporation takes over the result oriented responsibilities, nothing can be said in its praise. There has to be a radical change from conservation forestry practices to a more dynamic programme of production forestry (including man made forestry). The Forest Department should take initiative in transferring such areas to the Corporation which need reforestation and development.

A great advantage that has accrued to the Forest Deptt with the taking over of the forest exploitation by the Corporation is that the Forest Department has been saved from a great botheration of realization of Government dues from the Contractors. It has even saved the department from expenditure which it had to incur in the courts for the realization of dues.

As regards the training of the logging personnel much is left to be done. The labour to be engaged in logging should be recruited on a permanent basis and trained in modern lo the price of timber. It is even alleged that the forest Corporation has been responsible for the undue rise in timber prices due to perhaps under-utilization of felled trees and escalation of working expenses and non-prevention of losses coupled with heavy overhead (administrative expenses) of the Forest Corporation like any other corporation. This single factor has made the Forest Corporation unpopular with the general public which expects timber and fuel at a cheaper rate from the Forest Corpn.

The system of recording and checking at various stages is very strict and elaborate, but knowledgeable persons harbour some fear about pilferage for once a 'ravannah' has been prepared and issued and the material loaded, checking at any intermediate stage is just a formality, being merely as ticking process. This pilferage can not be detected until and unless the whole inventory of the Depot is prepared, and which it has not been possible to do so, it is reliably learnt. The recent big fire in the Corporation Depot at Lalkua is suspected to be a deliberate attempt to prevent detection of any shortage or discrepancy.

The Present Marketing System of the Forest Corporation

The present marketing policy of the U.P. Forest Corporation is that as far as possible the supply of forest produce be made to the consumers directly and priority be given to the Government Department, public sector undertakings and other public concerns. But during 1985-86, only 25.6 per cent could be made available to the consumers by direct sale.

The main reason why the marketing policy could not be fully implemented is that so far the Forest Corporation does not have any such surveying cell which could ascertain the main consumers of timber through a survey of the different markets for various kinds of forest produce. It is worthy of note that with a view to develop marketing organization in the forest Corporation, appointment has been made of a Marketing Manager, who is making efforts in this direction.

At present, majority of depots of the Corporation are situated near the forests. From 1985-86, the Corporation is setting up new depots in timber 'mandies', keeping in view the convenience of the consumers.

Only 18 saw-mills are being run by the Forest Corporation at present. This number is to be extended and in time to come sufficient number of saw-mills will have to be set up for sawing of as much timber as is mostly required by the consumers.

Marketing in Forest Corporation is different from private sector marketing. In private sectors, there is keen competition, so the purchaser gets advantage of competitive price and also gets the products of his choice. On the other hand the Forest Corporation enjoys a monopoly and faces no competition. So, in this state monopoly of forest produce trade, the general public is deprived of competitive, thereby cheaper price, and is also unable to get the quality product of his choice and of specified dimensions. One is in a way, forced to buy whatever is offered and from whatever depot it is available at a price which is monopolistic with added cost of transport from unwanted place.

Channels of Distribution : There are only one hundred six sale depots in the whole of Uttar Pradesh at selected places as shown in the Table. These depots do not cover all the regions. They also do not cover any region evenly. Moreover, these depots do not contain all kinds of produce. In the depot at Agra, which is such an important trade centre, none of the important hardwood was ever found present. Besides these depots, there is no other channel of distribution, so the public at large has ultimately to depend upon some middleman, agent or retailer in the open market. The system of getting timber by private treaty directly from the Divisional Logging Manager is so tedious that buyer can hardly avail of it. However, the opening of new depots in the timber mandies will solve this problem to some extent.

Product Development and Planning : The Forest Corporation has so far been dealing with only the primary products in a few species. No attention has been paid towards the processing of secondary products or their bye-products. This has delimited the market of the Forest Corporation. So, there is a need for planning in this direction, otherwise the Forest Corporation Sale depots will remain only as a centre of departmental supplies. There is not the least doubt that the country is going to face a timber famine. The practices of use of timber are prodigal and wasteful. A more rational and disciplined use of timber is the need of the day. Towards this end, it would be necessary for the Corporation to mount an aggressive campaign for the promotion of secondary timber species through seasoning and or treatment.

For this purpose, a chain of seasoning and treatment plants should be set up by the Corporation. These treatments are known to prolong the life of timbers and this would naturally result in adding value to the timber and in a reduced use of wood of scarce or valuable species.

Advertising : Whereas in some Corporations, like M.P. Forest Corporation, there is enough provision of publicity and frequent advertisements are seen in the news papers, there is no such advertising of sales etc. by the U.P. Forest Corporation with the result that the general public remains unaware of the sales by the Corporation. As a consequence of this, the same old trader, agent or middleman keep on purchasing timber from the Corporation sale-depots and later on selling them in the market at a higher rate. There is thus an imperative need for suitable media for advertising and publicity for which some budget should be set apart. A competitive culture shall have to be developed.

Market Research : At the same time the requirements of the market should be ascertained both with regard to the species and the sizes and dimensions preferred. This should in term be conveyed to the Logging Incharge to enable him to cut logs in suitable sizes in the forest itself. Thereby saving consequent wastage in off-cuts. Simultaneously the requirement of the secondary products can also ascertained. The need for a separate market survey cell is evident.

Salesmanship : In a timber market, one is apt to loose sight of good salesmanship but a goods salesman is very necessary if the Corporation does not want to lose a customer. There are some cities like Delhi, Agra, etc. where timber of more than one Forest Corporation is stocked and if the market principles of salesmanship are not applied, sales are likely to be adversely affected. The Corporation depots are mostly in charge of a Forester or a Deputy Ranger, who are used to rough dealing in the forest. They need to be assisted by suitably selected and trained salesman. Such salesman can replace existing scalers on daily wages and will therefore, not entail much extra expenditure.

Aggressive Marketing Strategy : The Forest Corporation can adopt aggressive marketing strategies for identifying customers in advance and gearing production through advance supply orders. With proper integration, wastage in use of wood is being minimised.

A further contribution to integration is the setting up of saw milling-cum-seasoning-cum-treatment-cum-joinery units. The establishment of such units will a long way in promoting the use of seasoning timbers in place of conventional timbers like Teak, Sal, Deodar, etc.

It needs to be ensured that the Corporation follows an aggressive marketing policy to enable them to build healthy order book position. The package of services provided by the Corporation should be such as to attract consumers from far and wide. There is a great need for personalised attention so far as consumer services are concerned.

Public Relations : is an important activity of any concern or undertaking. It is a necessary adjunct in Govt. Departments also. This is necessary to coordinate the sales with production and allied matters. So, either public relations personnel be appointed separately or some of the existing staff members be assigned the task of public relations after suitable training.

After Sales Service : is not of any relevance in the timber trade for the present, but as and when the secondary products come in the market, an eye will have to be kept on the durability of such products. Such study may help in improving the quality of produce.

Transportation : Some depots are situated at out of the way places far away from towns and there are no facilities for transport of timber as a consequence of which general public, which is a small buyer can not purchase timber from such depots because it can not economically transport timber to his place. So again the refrain of emphasis comes on the establishment of more sale depots and near the towns. Otherwise, only the bulk purchasers (i.e. Traders buy from such outlying depots and sell in the market at great profit. Thus, the general public is deprived of timber supply at cheap rates which is one of the object of the Forest Corporation. Now the Forest Corporation has opened new depots in the timber mandies. This would go a long way in solving the aforesaid problems of the consumers.

Storage, Stacking and Warehousing : For proper marketing, storage and warehousing facilities are sine quanon. Unfortunately, the U.P. Forest Corporation depots are not good for storage. Valuable timber is stacked in the open with the result that the timber gets rotten and decayed (Photo No. 29) or catches fire as happened in Lalkua depot, where fire extinguishing measures were not at all available.

The Corporation should build up proper storage and ware-housing where timber may be kept away from direct sun or rain and the place should be kept safe from any fire danger. Such places should also be insured against fire.

Socio-Economic--Security And Effect

The forest resources are fast dwindling but the demand for timber and soft woods for paper is likely to increase further, so there is urgent need to preserve whatever timber we have. In fact, we should now insist on timber management in addition to forest management and curtail our fellings accordingly, even if goes against the working plan prescriptions. And in the timber depots also, the sales should be so organized that every bit of timber is sold and no one is allowed to carry timber to be stocked in the godown for subsequent sale at higher prices to poor public.

A few species should be suitably demarcated i.e. their marking and felling restricted so that these can be made available insubsequent years in a sustained manner.

As is evident, there are not many retail depots and some of them are far removed from consumer centres, so these become the monopoly purchase centres for the stockists who indulge in black marketing. This can be checked by opening as many retail centres as possible and by making controlled sales and also by making general public aware of existence of such sale depots through advertisement and publicity.

Raw Material of Forest Based Industries : Mention has already been made in the previous chapter about the efforts made by the Corporation to establish forest based industries. Any blueprint for a future strategy of the Forest Corporation should give a high place to this topic and endeavours should be made by the Forest Corporation to diversify their activities and to take in their fold industrial development over a wide spectrum, which will embrace the tiny, cottage scale, small scale and medium and large sectors. In all industrial planning, the key word should be integration and full utilization of raw-materials. It will be difficult in the earlier stages for the Corporation to tap the requisite expertise required for establishing industries but once a beginning is made, there should be no problem in future and with experience and sufficient expertise and confidence would be available.

Plantation And Afforestation

National Commission on Agriculture had recommended that the Forest Corporation should confine plantation activities to potential forest areas carrying a forest crop with low value. There has been a good deal of criticism of this idea in the State Forest Department. There is a growing feeling that the Forest Corporation must also take up problem forest areas or degraded forests in order to make them productive.

There is a proposal for the preparation of a Corporate plan for the Forest Corporation during the next 10 years. Under this plan, the Corporation will undertake plantation of industrially important species like Eucalyptus.

The Forest Corporation can do a yeoman's service by encouraging private forestry as well. This is a new concept for the country and, if successfully implemented, it can result in making the country green. There are excellent examples of similar ideas being practised in Western countries.

Racket in forest trade : Just as in other trades, so in the forest trade also, there are several rackets, some of which are made possible due to faulty organization and others are due to bad mentality of corruption prevalent in the society.

The sale depots are generally manned by officials of lower rank but they are vested with wide powers and given charge of forest produce of great value. Similarly, the scalers who are said to be daily wages workers are entrusted with the task of measuring the timber sold and it is on these measurements that the bills of payment by contractors (purchasers) are prepared. Naturally, there is great opportunity for the low paid scalers to indulge in corrupt practice as has been alleged on various occasions.

Some kind of racket is also reported to prevail in the matter of grading of timber lots and changed of grading labels after sales with ulterior motives. This kind of racket, it is feared leads to a fall in the sale price of timber lots, because the purchaser, not feeling sure of the receipt of the highest grade, bids considering the lot to be of average or lowest grade.

Problem of Finance (Credit Control, Investment Appraisal)

Owing to its being a local authority, the financial structure of the Forest Corporation rests on loans received from the Govt. and cash credit facilities available from certain banks.The Forest Corporation is not authorised to collect capital and therefore, in the initial years it faced great difficulties in procuring finance. At present, cash credit facility to the tune of Rs. 295 lacs is available to the Corporation from Bank of Baroda. The Forest Corporation is reported not to have made use of cash credit facility from 1982 - 83 to date. Till the end of 1985 - 86 (dated 30.9.86), the own capital of the Corporation is estimated as Rs.11.75 crores.

To enable the Forest Corporation to obtain long term loan at cheap rate for development works, proposal is under consideration for its reconstitution and getting it registered under the Company Act.

Law And Order Problems

The Forest Corporation also faces certain law and order problems. There have been cases of forcible lifting of timber and other forest produce from the forests coupes and the Corporation's sale depots by unsocial elements on the point of gun.

During 1984-85, occurred several cases of burning of tendu leaves stacks and lotting of tendu leaves from the stacks and trucks. There has been sharp decline of such cases during 1985-86. Some theft also occurs while the timber is being floated down the rivers.

The traditional labour from certain places does not move out of their homes without taking sufficient advance for the family. Sometimes, this advance is not fully recovered due to some unforeseen reasons, so the balance has to be written of causing some loss to the Corporation.

Tax Problems

Being a local authority, the U.P. Forest Corporation should be exempt from the payment of 'Corporate Tax'. But the Income Tax Department is not prepared to give this exemption, so this matter is subjudice in various courts.

The decision of the Sales Tax Department to charge Sales Tax on the whole turn over of the Forest Corporation instead of Tax on the sale price fixed by the Forest Department at the point of transfer is under challenge in the High Court.

In addition, the matter of demand of "Mandi" Tax by the U.P. Mandi Parishad without affording any services and facilities to the Corporation is pending in the Supreme Court. Similarly, matter of imposition of central production tax on sawn timber of the Corporation by the Central Government is under challenge in the High Court, because of its impropriety.

Staffing Problems

The staffing needs of the Corporation for forestry personnel are quite heavy. Initially, as per recommendations of the National Commission on Agriculture, staff was obtained from the Forest Department on deputation. While this was a good idea, making for a two-way flow of experience and expertise, it started putting a strain on the forest department on account of a darth of trained staff. From 1982 - 83, there has been a sudden increase in the work load of Corporation for which suitable staff could not be had from the forest department in adequate number and thus the Corporation felt great hardship in proper working. Since then there has been some improvement, but the problem still looms large. This experience supports the need for a sizeable element of open market recruitment for the Corporation and for suitable training facilities for their staff, for which the Corporation may have to consider opening their own training centres for extending the facilities and benefits of training to all the levels of employment. It is heartening that the Government of India has set up an institute of Forest Management at Bhopal in collaboration with the Indian Institute of Management, Ahmedabad for training forest and enterprise managers in the techniques of business management.

At the same time, the problem of overstaffing or excessive man power in some segments of the undertaking also requires to be resolved.

Suggestions For Costing Wing

With the increasing volume of work in the Corporation it would be necessary to streamline costing and accounting procedure. For this purpose, the Corporation would do well to set up a strong costing wing which could be entrusted with the task of preparing realistic budgets, fixing norms, carrying out variance analysis, involving standard costing procedure, carrying out pricing studies and undertaking monitoring and appraisal work.

Appraisal And Evaluation

In any Public Sector Organization, there is need to carry out regular performance appraisals. In the Central Government, there is an apex body known as Bureau of Public Enterprises, which looks after central public sector corporations, apart from playing the role of a friend, philosopher and guide. The U.P. Forest Corporation has entrusted this task of appraisal and evaluation to the U.P. Development System Corporation Ltd., Lucknow which is a State Government Undertaking. It is believed that this organization has failed to evaluate the Forest Corporation.

Engagement of Educated Unemployed

The most important object great social value was the engagement in forestry work of the education unemployed youth after giving them training in modern logging technique. But unfortunately precious little has been done in this direction. Enquiries from concerned authorities could not reveal any plausible reasons for not undertaking this important item even 14 years after the inception of the Corporation. Had a beginning been made in this direction, the paucity of staff which is so acutely being felt now would not have arisen. This object should now be persued in right earnest diligently and seriously otherwise the wrath of the educated unemployed youngmen and women may disturb the performance of the Forest Corporation.

Construction of Roads

The U.P. Forest Corporation was also entrusted with the responsibility of constructing the roads for working in the hilly areas and far off inaccessible regions. This was to be performed even much before taking up the extraction work in the forests. Such work could never be done by individual contractors, so in the emergence of the Forest Corporation people saw bright prospects of such road construction which would benefit the local population besides serving as the export roads of forest produce. It appears that this objective is being assigned the last priority which is detrimental to the progress of the Corporation and the nation, both.

A serious loophole which has escaped attention is the lack of adequate supervision and checking in the Corporation working. In the former system of working, because every one was suspicious about contractors, so there was strict checking by the forest officials from the lowest (stump counter) to the highest level and even the neighbouring villagers kept a vigilent eye, but now everyone is complacent thinking that the Forest Corporation is State owned, so there should not be any room for any wrong or illicit practice in its working. The forest department officials too deem it unnecessary to do any serious supervision or checking. Rather they had been cooling their heals since the Forest Corporation stepped in the field of forest exploitation. Consequently, some malpractices have crept into which are similar to illicit fellings in different garb, and these may assume serious proportions, if proper steps are not taken to ship them in the bud.

To remedy this, it is suggested that a panel of forest officers (including a retired officer) and representative of the public and the former contractors should be made to inspect the forest operations undertaken by the Corporation. Thus, a true picture of the situation really prevailing would emerge and the alleged wrong practices would automatically cease.

In the early years, the Forest Secretary to the Govt. of U.P. remained the Chairman of the Corporation. On the one hand this helped in the establishment of the Forest Corporation but on the other, it gave certain undue latitudes to the Forest Corporation at the cost of the Forest Department. The formula for the calculation of royalty which has been involved is biased in favour of the Corporation and deprives the Forest Department of proper royalty. In this formula, a factor should be introduced which may incorporate the ever increasing price of the timber in the market so that proper royalty may be paid to the Forest Department.

While discussing the various expects of processing and marketing and the performance of the U.P. Forest Corporation, suggestions for improvement have been given side by side, yet these are being summarised below :

Suggestions

(1) Introduction of better and stricter business management attitudes, whereby forest managers are learning to treat activities as task oriented, time bound, cost effective and cost contain.

(2) Realization of the importance of the accountability and financial discipline.

3) The remarks of the Prime Minister (The Hindustan Times, dated December 23, 1987) that the public sector was eating up the wealth of the nation instead of generating it, should be born in mind by the Forest Corporation also. The Forest Corporation should avoid huge offices and unnecessary paraphernalia. The Corporation should function to generate forest wealth, thereby national wealth and should plug all holes of dissipation of country's most important natural resource. Expenditures on administration should be reduced and austerity should be observed in the use of vehicles, by the Forest Corporation Officials.

(4) Research and development should be expedited specially in mensuration and yield studies, so that accurate estimates of out turn be prepared and under utilization of trees be stopped.

(5) The Corporation should go in for complete modernization of old methods of forest extraction, and total mechanization of various operations.

(6) The Corporation should encourage the development of labour cooperative societies and should also make proper use of the already existing ones.

(7) Supervision by the Forest Department should not be done away with merely because now the felling etc. are being done by State owned Corporation.

(8) Fire safety measures must properly be provided both : in the forest coupes and in the sale depots to avoid fire hazards.

(9) Insurance against fire must be got done both in the forest and the sale depot.

(10) Social justice to labour in the form of security, insurance, etc. should be provided. Labour should be made permanent after specified number of years as is the case in Railways and other Central Government Departments.

(11) The Corporation should extend it marketing activities to remote an inaccessible areas.

(12) Construction of roads in the hilly regions and inaccessible areas must be expedited, so that traditional method of carriage may be replaced by modern transport systems.

(13) The inventory of forest produce expected and present in the depots must be prepared regularly and verification should be carried out periodically so that pilferage, if any, may be detected timely.

(14) The Corporation should take proper care in marketing the produce so that the timber which is getting spoiled because of long storing time may be sold first.

(15) The Institution of Scalers must be replaced by responsible and well paid permanent personnel.

(16) The sales depots should also be manned by officials of higher rank.

(17) Just as in Sweden and other countries, trucks for transport should be taken on hire instead of purchasing departmental trucks, so that expenditure during idling of trucks is avoided.

(18) The Corporation should go for mix sellings so that the wood of inferior species could also be sold at competitive and higher price. For this, the Corporation should sell inferior wood alongwith superior species so that higher prices may be fetched.

(19) The Corporation should also go for controlled sales or even rationing of those species which are short in supply and in place of these species, the Corporation should encourage other species after seasoning.

(20) Last but not least, is that the Forest Corporation, like any other public sectors undertaking should not run like Government Departments.

On the whole, the U,P. Forest Corporation has laid the foundation of sound and integrated forest management. It should now also embark upon timber management for the people through its specialized logging and marketing wings.

Bibliography

Agarwal, V.P. - Introduction of Mechanised Logging in Himachal Pradesh, 11th S.C., 1968, F.R.I., Dehradun.

A Brief History of Forest Administration in U.P., The Centenary of Forest Administration in U.P., 1861-1961, U.P. Forest Deptt., 1961.

Chauhan, J.S. - Forest Resources in Kumaon Region of Uttar Pradesh and their exploration, (Unpublished), 1972.

Chauhan, J.S. - Forest Resources in Kumaon Region of Uttar Pradesh and their exploitation (Unpublished), 1975.

Clawson Marion - Economics of National Forest Management.

Champion, H.G. & Seth, S.K. - Forest Types of India, Government of India (Revised Classification), 1968.

Das, E.S. - Extraction methods in Western Himalayas, with S.C., F.R.I., Dehradun, 1968.

Day, H.C. - Forest Administration in Kumaon, The Centenary of Forest Administration in U.P. - 1861-1961, U.P. Forest Deptt., 1961.

Diwedi, A.P. - Forestry in India.

Duerr, W.A. - Fundamentals of Forestry Economics, McGraw Hill Book Co., 1960.

Encyclopaedia Britannica, Articles on Forestry.

F.A.O. (1974) - An Introduction to Planning Forestry Development, Rome.

F.A.O - Planning a Forest Inventory.

Forest Statistics, Uttar Pradesh, upto 1982, U.P. Forest Deptt. R & D Circle, Lucknow.

Forest Statistics, 1983, U.P. Forest Deptt., R & D. Cir., Lucknow.

Forest Statistics, 1984, U.P. Forest Deptt., R & D. Cir., Lucknow.

Forest Statistics, 1985, U.P. Forest Deptt., R & D. Cir., Lucknow.

Forestry in India - A Sector Study, Swedish University of Agriculture and Forestry, Stockholm.

F.R.I. & States Annual Research Reports.

F.R.I & College - Indian Forest Utilization, Vol. I. & II, Editorial Board, F.R.I., Dehradun.

F.R.I & College - The Indian Foresters.

Gane, M. (1966) - Forest Harvesting, Common Wealth Forest Review.

Ganguli, B.N. - Analysis of cost and methods of Forest exploitation, 11th S.C. F.R.I., Dehradun, 1968.

Govt. of India - Report of the National Commission on Agriculture Part IX, Forestry, 1976.

Govt. of India - Publications of Forestry Cell, Planning Commission.

Govt. of India - Publication of Ministry of Agriculture and Irrigation.

Govt. of India - India's Forests 1980, Central Forestry Comm.

Govt. of India - Report of the Task Force on Forest Resource Survey. Planning Commission, New Delhi.

Govt. of India - India's Forests 1976, Central Forestry Comm.

Govt. of India - Indian Forestry, Forestry Division, Dept. of Agriculture, Ministry of Agriculture, N.Delhi.

Govt. of India - Forest Statistics Bulletins for various years, Central Forestry Commission, New Delhi.

Gulati, N.K. - A Glossary of Forestry Terms, International Book Distributors, Dehradun, 1981.

Hall, W.T. - The Centenary of Forest Administration in U.P., 1861 - 1961, U.P. Forest Deptt., 1961.

Haltzclaw - Principle of Marketing.

Hundred Years of Indian Forestry (1861-1961) - F.R.I. and College, Dehradun, 1961, Centenary Celebration Vol. I & II.

Johnston, D.R. Grayson, A.J. & Bradky, R.T. - Forest Planning; Faber & Faber Ltd., 1967.

Joshi, D.P. - Logging Techniques through the Century. The Centenary of Forest Administration in U.P., 1861 - 1961, U.P. Forest Deptt., 1961.

Joshi, P. - Distribution of Medicinal Plants in Hill districts of U.P.

Kant, Hari - Wood harvesting for fuel wood, charcoal and Small Scale Industries, Andre Mayer Fellowship Report, F.A.O., 1980.

Khandawala, P.N. - Design of Organization ; Harcourt Brace Jovanovich International, New York, 1977.

Kotler Philip - Marketing Management : Analysis, Planning & Control; EEE, Prentice Hall of India Pvt. Ltd., 4th Ed. 1980.

Maslekar, A.R. - A study of Auction as a Marketing Mechanism for Forest Products. Ph.D. Thesis (Unpublished) Gujarat University, 1981.

Do - - A Study of Marketing of Wood Products in Maharashtra, I.I.F.M., Monograph No.1, 1980.

Mazur, Paul - The Standards we raise, Harper & Row, New York, 1953.

Mehta, S.C. - Marketing Environment, Concept and Cases ; Tata McGraw Hill Publishing Co. Ltd., 1980.

Mehta, S.C. & Maslekar, A.R.- Sales Depot at Delhi : A new approach to timber Marketing; Case IIMA/IIFM, 1979.

Mehta. T. - Lecture Notes on Forest Utilization, Part I & II for Forest Ranger students.

- Do - - A Hand Book of Forest Utilization, Periodical Expert Book Agency, New Delhi, 1981.

Mohapatra, C.R. - The Forest Corporation - A new venture in Indian Forestry. 8th World Forestry Congress, Jakarta, 1978.

Padhi, G.S. - Forests of India - A Critical Study International Books Distributors, Dehradun, 1982.

Padhi, G.S. & Padhi, A.K. - Hand Book of Forestry, 1975.

Pal, I. - Forest Wealth of Uttar Pradesh, The Geographer Vol. III.

Pearson, R.S. - Commercial Guide to the Forest Economic Product in India.

Published and unpublished data and material from Addl. Chief Conservator of Forests (Management).

Published and unpublished data and material from Managing Director, U.P. Forest Corporation.

Publications of Foreign Countries on Forests and Forest Utilization.

Ralph, S.A. - Report of the Definition Committee, Journal of Marketing, October 1948.

Report on Northern Regional Seminar, Nainital, 1977, U.P. Forest Department.

Sagreiya, K.P. - Forest and Forestry, N.B.T. India, N. Delhi, 1967.

Sahai, S.P. - Forests - Their Geographic & Economic influence, The Centenary of Forest Administration in U.P., 1861-1961, U.P. Forest Dept., 1961.

Seth, S.K. - Medicinal Plants, N.B.T., India, N.Delhi, 1968.

- Do - - Forest Management & Working Plan. The Centenary of Forest Administration in U.P., 186-1961, U.P. Forest Department, 1961.

Seth, S.K. & Lohani, D.N. - One hundred years of Forestry in Uttar Pradesh. The Centenary of Forest Administration in U.P., 1861 - 1961, U.P. Forest Department, 1961.

Seth, V.K. - Forestry Corporation in India : Present status and future strategies. 2nd Forestry Conference, F.R.I., Dehradun, 1979.

Sharma, L.C. - Forest Economics, Planning & Management; Bishen Singh Mahendra Pal Singh, Dehradun, 1980.

- Do - - The Forest Industry in Uttar Pradesh - Ph.D. Thesis, Agra University (1972).

Singh, J.B. - Marketing of Forest Products; Problems & Prospects. Seminar on Forestry and Forest based Industries; N.P.C., April 1980.

Singh, G.N. - Forest Administration & Public opinion. The Centenary of Forest Administration in U.P. 1861 - 1961, U.P. Forest Deptt., 1961.

Sahai, R. - The Forest of Uttar Pradesh & their importance, The Centenary of Forest Administration in U.P. 1861 - 1961, U.P. Forest Deptt., 1961.

Sisam, J.W.B. - Teaching Forestry & Utilization - A basic curriculum for Universities in developing countries, Unasylva, 1964.

Soni, R.C. - Forests of U.P. The Centenary of Forest Administration in U.P., 1861 - 1961, U.P. Forest Deptt., 1961.

Stebbing, E.P. - Forest utilization in India, Vol. III & IV, F.R.I., Dehradun.

- Do - - The Forests of India, 3 Volumes - London.

Thapar, S.D. - India's Forest Resources. McMillan & Co. Ltd., India.

Thomas, P.S. - Study of Forest Development Corporations in India (IIAM/IIFM), 1978.

Tiwari, D.N. - State Trading in Forest Produce in India; Jugal Kishore & Co., Dehradun, 1981.

Trotter, W.H. - Manual of Indian Forest utilization, 1940, F.R.I., Dehradun.

- Do - - The Common Commercial Timbers of India and their uses (1940).

Troup, R.S. - Indian Forest utilization, F.R.I., Dehradun.

Troup. R.S. - Silviculture Systems, 1952, The Clarendon Press, Oxford.

U.P. Forest Corporation - Act, 1974, U.P. Govt., 1975.

U.P. Forest Corporation - A Short Note on its activities, 1974 - 75 to 1984 - 85.

U.P. Forest Corporation - Annual Report 1975 - 76.

U.P. Forest Corporation - Annual Report 1976 - 77.

U.P. Forest Corporation - Annual Report 1977 - 78.

U.P. Forest Corporation - Annual Report 1978 - 79.
U.P. Forest Corporation - Annual Report 1979 - 80.
U.P. Forest Corporation - Annual Report 1980 - 81.
U.P. Forest Corporation - Annual Report 1981 - 82.
U.P. Forest Corporation - Annual Report 1982 - 83.
U.P. Forest Corporation - Annual Report 1983 - 84.
U.P. Forest Corporation - Annual Report 1984 - 85.
U.P. Forest Corporation - Annual Report 1985 - 86.
U.P. Forest Department - General Conditions of Sale.
U.P. Forest Department - R & D. Circle, Lucknow - Raw Material Resources of U.P. for Wood based industries, 1968.

INDEX

V

W

Y

Z